Stereochemistry at a Glance

Jason Eames
BSc (Sheffield), PhD (Cambridge)
Lecturer in Organic Chemistry
Department of Chemistry
Queen Mary, University of London
and Lecturer in Organic Chemistry
St. Hilda's College, University of Oxford

Josephine Peach
BSc, MA, DPhil (Oxon)
Fellow and Tutor in Chemistry
Somerville College
University of Oxford

Blackwell
Science

First published 2003

Library of Congress
Cataloging-in-Publication Data

Eames, Jason.
  Stereochemistry at a glance / Jason Eames, Josephine Peach
    p.   cm. – (At a glance series)
  Includes index.
  ISBN 0–632–05375–5 (softcover)
  1. Stereochemistry.   I. Peach, Josephine.   II. Title.
  III. At a glance series (Oxford, England)

  QD481.E25 2003
  547′.1223–dc21

                                              2003048169

ISBN 0–632–05375–5

A catalogue record for this title is
available from the British Library

Set in 9.5/11.5 pt Times
by Integra Software Services Pvt Ltd, Pondicherry, India

For further information on
Blackwell Publishing, visit our website:
www.blackwellpublishing.com

FSC
Mixed Sources
Product group from well-managed
forests and other controlled sources
Cert no. SGS-COC-2953
www.fsc.org
© 1996 Forest Stewardship Council

Contents

Symbols and abbreviations can be found on the inside back cover

Introduction

This book aims to promote the understanding of stereochemistry in organic chemistry (and thus its use in organic synthesis in general) by following a systematic approach to the solution of problems based, as in its sister volume by Mark Moloney, on deductive reasoning. The importance of good drawings is emphasised and explanations based on simple molecular orbital theory are used, alongside more conventional answers, to underline similarities between reactions. The text is aimed primarily at first and second year university students, but a brief general introduction to stereochemistry is provided in Chapter 1 for those who may have only a limited knowledge of the topic. In subsequent chapters, the same general strategy is applied to every question. Hints and comments are available at each stage, allowing progress to be made and assessed. This format gives students the option not only to check their final answers, but also shows them how to dissect, start and progress within a problem. From our experience, the analytical stage of problem solving is often neglected. We hope that users will gain not only a logical approach to the solution of problems, but also an understanding of the shape, symmetry and stereochemistry of organic molecules.

Acknowledgements

We are most grateful to all of those who have helped us in the preparation of this book, especially for the constructive advice and comments from Leonie Blackburn, Dr Stephen Faulkner, Stephanos Ghilagaber, Michael Suggate and Dr Michael Watkinson. To all of these we record our sincere thanks. In addition we would like to thank Dr Mark Moloney for his advice, help and many interesting discussions concerning the teaching of organic chemistry.

We would like to dedicate this book to Jennifer Eames and to John Peach for their continuous support.

How to Use This Book

Read over Chapter 1 if your knowledge of stereochemistry and its technical terms need refreshing. Do the questions at the end of Chapter 1.

The remaining chapters each has a short introduction and then a series of questions and answers.

Some dos and don'ts

- DO read the question very carefully!

- DO write down your answers on paper

- DO check your answers at the end

- DO cool down, calm down and approach problems logically

- DON'T guess the question from keywords

- DON'T uncover hints unless really needed

- DON'T force an answer to fit if it does not

- DON'T panic!

How to do the questions

- Cover up the right-hand page (**E**, **F**, **G** and **H**) with a sheet of A4 paper and use this for your answers. Read the question **A**.

- Start to answer it using the strategy outlined in the first stage in **B**; draw out the structures and make deductions from them. If you are struggling, read the corresponding comments in **C**. Further help is available in **E**.

- Go on to the second stage, looking at the reagents; make deductions and suggest a possible mechanism. Use the comments **C** to help you along if you are stuck, or even more help at **F** and **G**.

- Work similarly through the third stage, which involves drawing out a possible answer, making sure that it answers the question that is asked and fits all the data given.

- Summarise your answer. Finally uncover **H** and compare your answer with **H** and **D**. They will not be identical, but the same points should be made and the same conclusions should be easily reached by another student using your drawings.

Page layout for the questions

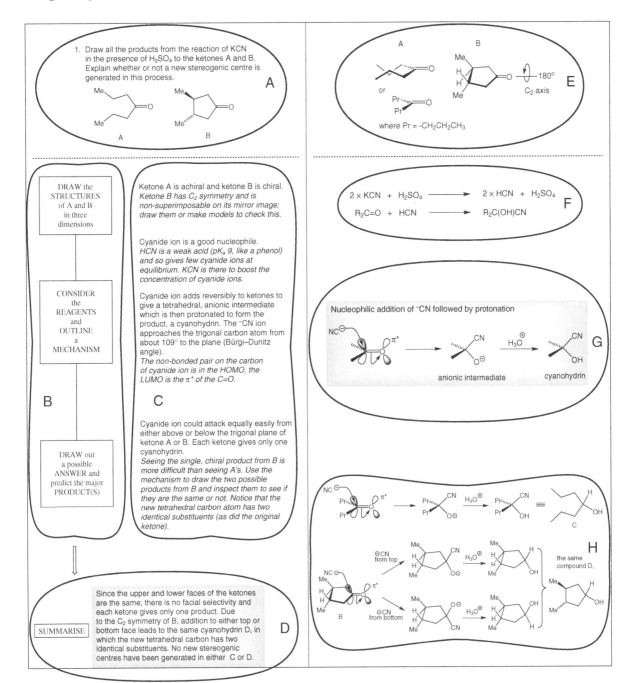

LEFT-HAND PAGE

A: the question

B: three-stage strategy for answering

C: deductions and comments for each strategy stage; comments in italics include additional explanations or further comments

D: a written summary of the answer

RIGHT-HAND PAGE

E: drawings and deductions from the structures given in the question

F: general equation(s) for the reaction(s) concerned

G: general mechanisms, showing stereochemistry with the orbitals involved

H: drawings for the answer

*An ideal answer might be a combination of parts **D** and **H***

1 Shapes and Orbitals

Stereochemistry is about the three-dimensional shapes of molecules and how these affect or control their reactivity. The shapes of organic species are a consequence of the spatial arrangement of electrons in their orbitals – electrons bonding to another atom or non-bonding electrons (lone pairs). Every bonding molecular orbital has a corresponding unoccupied, antibonding molecular orbital. Antibonding orbitals are as important in reactions as bonding ones.

Most organic reactions occur between an electron-rich species (a nucleophile) and an electron-deficient species (an electrophile), which interact initially through the highest-energy occupied molecular orbital (HOMO) of the nucleophile and the lowest-energy unoccupied molecular orbital (LUMO) of the electrophile. It is crucial to know where these orbitals are positioned in space in order to predict reaction pathways.

Even after we have worked out the three-dimensional shape of a molecule and the spatial arrangement of its orbitals, we still need to be able to represent these on paper in only two dimensions. Conversely, we must also be able to take a two-dimensional picture and imagine it, in the mind's eye, in three dimensions. This can be difficult and often needs practice.

Drawing conventions

There are a number of conventions to help us to translate between two and three dimensions. The most important drawing convention is the use of solid, wedged and dashed lines for bonds.

Solid lines	——	mean that the bond is *in the plane of the paper*.
Wedged lines	◤	mean that the bond is *coming out towards you*, in front of the plane of the paper, with the thicker end of the wedge closer to you.
Dashed lines	ⅲⅲⅲ	mean that the bond is *going away from you*, behind the plane of the paper. Don't confuse this with the dotted line - - - -, which we will use for partially made (or partially-broken) bonds in transition states.

Other drawing conventions are described on page 6.

Some students find that imagining a two-dimensional drawing in three dimensions is easy, but the majority find it difficult at first. If you don't find it easy, and many people do not, then there are two things to remember.

(1) PRACTISE. Keep trying, it really does get easier. Above all, *don't give up*, it matters.
(2) USE MOLECULAR MODELS. These help enormously.

The standard molecular model kits have hollow plastic bonds like lengths of drinking straws. A set is quite inexpensive, but make sure that you get a set suitable for organic chemistry; these have more tetrahedral atoms and fewer octahedral atoms and other exotic shapes.

Many students think that if they use molecular models they will become dependent on them, and that they will be lost without them. Experience has taught us that the reverse is true: the more you use models, the more familiar you become with the shapes, the *less* you need them. So use them as frequently as you wish, especially to check your own visualisation.

Models have the advantage that you can pick them up, turn them around and put them in any orientation to suit you. The next best thing is a good computer modelling program, with which you can build molecules in three-dimensional representations and rotate them in real time. Watching the rotation of a molecule is very helpful in visualising its three-dimensional shape.

Hybridisation: bonding orbitals for carbon

Three different major bonding arrangements are available for carbon, achieved by different mixing (by hybridisation) of its single 2s orbital with the three 2p orbitals ($2p_x$, $2p_y$ and $2p_z$). The hybrid sigma (σ) bonds are spread out as far as possible away from each other, with any pi (π) bonds as far away from the sigma (σ) bonds as possible and as far away from any other π-bonds as possible. This is a consequence of valence shell electron pair repulsion (VSEPR) theory.

	Orbitals mixed (for bonding)	Number of σ-bonds and spatial distribution	Number of π-bonds and spatial distribution	Example
sp³	one s and three p (for σ-bonds) no p left	4 Tetrahedral Bond angle ~ 109°	none	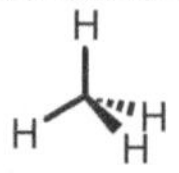 Methane, CH_4
sp²	one s and two p (for σ-bonds) leaving one p (for π)	3 Trigonal Bond angle ~ 120°	1 Perpendicular to σ-plane	 Ethene, $H_2C=CH_2$
sp	one s and one p (for σ-bonds) leaving two p (for π)	2 Linear Bond angle ~ 180°	2 Perpendicular to σ and to each other	 Ethyne, $HC\equiv CH$

Hybridisation: bonding and non-bonding orbitals for nitrogen and oxygen

Similar bonding arrangements are available for nitrogen and oxygen, except that for these one or two lone pairs (respectively) take the place of a sigma C–H bond. Non-bonding electrons are especially important as these are the HOMOs of nucleophiles.

	C example	N example	O example
sp³	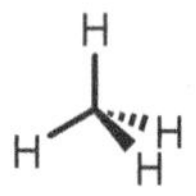 Methane, CH_4	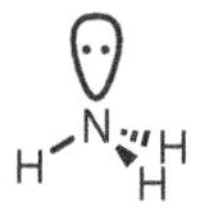 (PH₃ is similar)	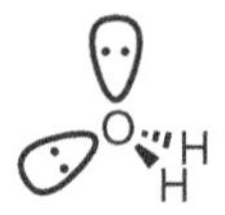 (H₂S is similar)
sp²	or	or	or
sp	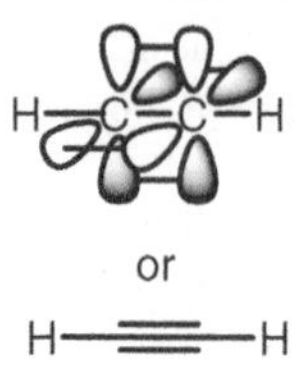 or $H\underline{\equiv}H$	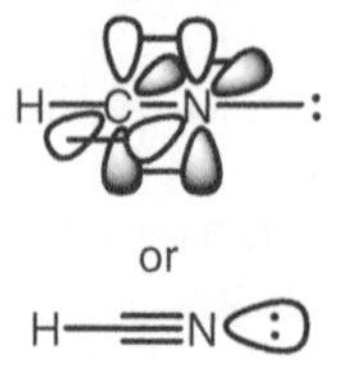 or $H-\equiv N$	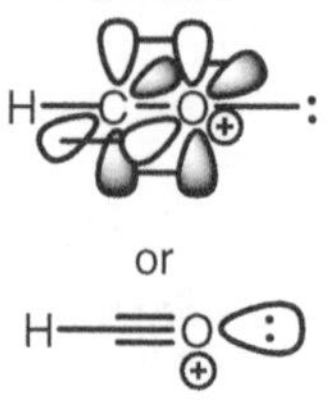 or $H-\equiv O$

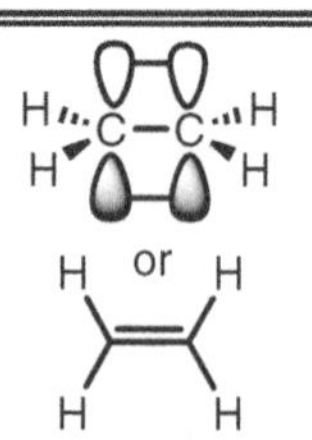

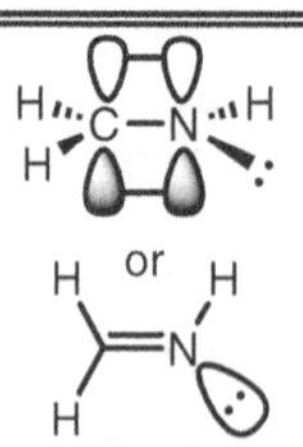

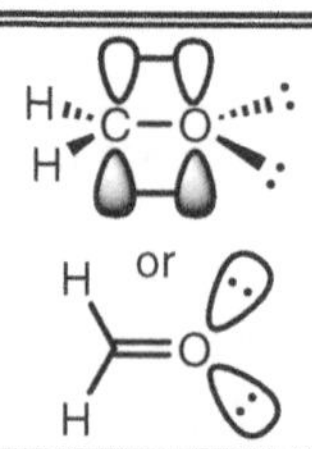

Antibonding orbitals

Every bonding orbital has a corresponding antibonding orbital, and these too have their own spatial arrangements. The σ* and π* are the most common LUMOs.

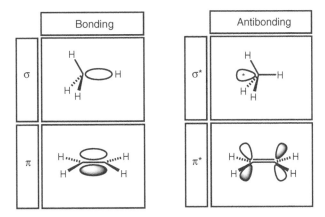

Energies of bonding, non-bonding and antibonding orbitals

The shapes and relative energies of the most important filled and empty orbitals are shown below. They are similar for the elements C, N and O.

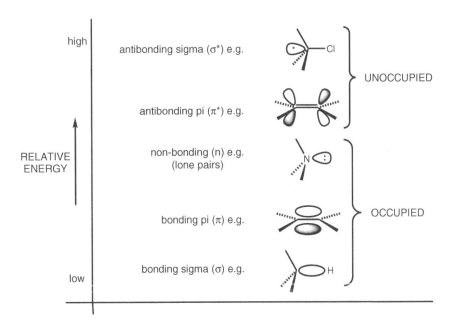

Nucleophiles and electrophiles

Polar organic reactions start with an interaction between a filled and an empty orbital, that is, between the highest occupied molecular orbital (HOMO) of the nucleophile and the lowest unoccupied molecular orbital (LUMO) of the electrophile. This interaction is most favourable when the HOMO and the LUMO are matched in size, shape and energy. Hence *the best nucleophiles have high-energy occupied molecular orbitals*, such as the non-bonding electrons of NH_3 or the π-electrons of alkenes and arenes.

The *best electrophiles have low-energy unoccupied molecular orbitals* such as the π* of alkenes or carbonyls, the σ* orbitals corresponding to polar σ-bonds, or the vacant p-orbitals of BF_3, $AlCl_3$ and a six-electron carbocation like Me_3C^+. It is essential to know where these orbitals are in space, as indicated above.

Drawing organic mechanisms

Curly (or curved) arrows are used to indicate the movement of electrons in organic mechanisms. *Don't* use curly arrows to show the movement of *atoms*. Curly arrows must start at a bonded or non-bonded pair of electrons (e.g. from a lone pair, σ- or π-bond) and end either where the new bond forms or on an atom. They must also take account of the stereochemistry of the orbitals concerned.

Starting with a very simple example: the ionisation of H–Cl:

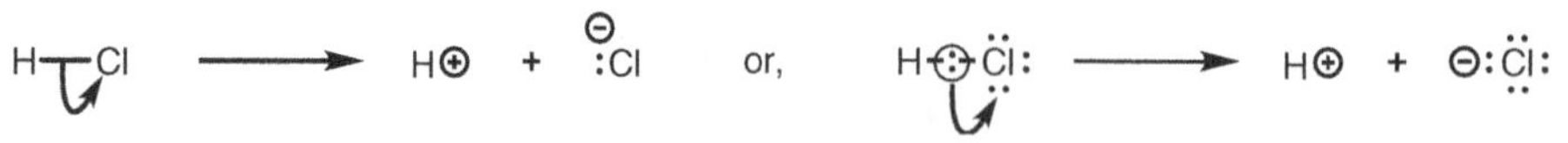

Equation 1

or, recombination of the two resulting ions.

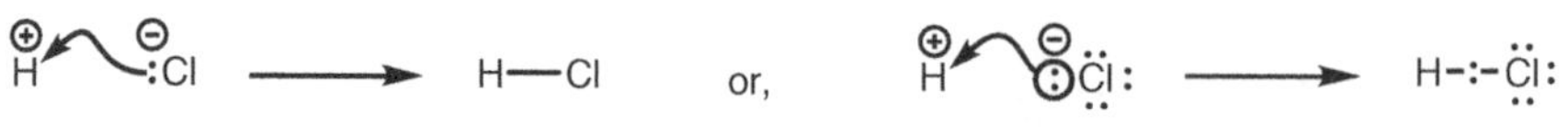

Equation 2

Generally, the non-bonded pair of electrons on a nucleophile is shown as a pair of dots, but not always – after a while, this will become second nature to you and it will be assumed that you know where these electrons reside!
Hence:

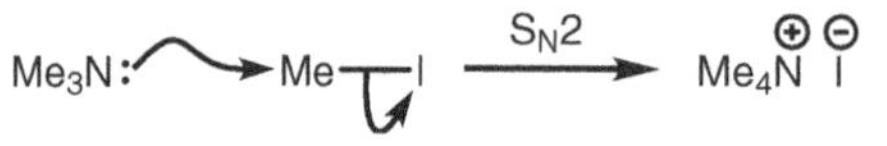

Equation 3

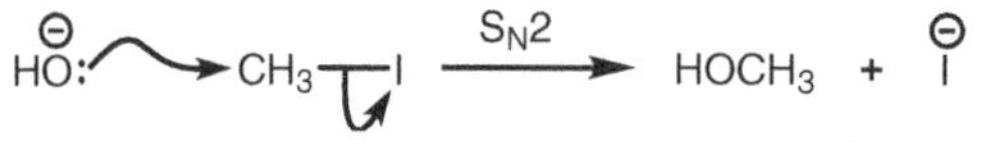

Equation 4

Several bonds may be broken or made at the same time in a concerted reaction, for example in an E2 elimination.

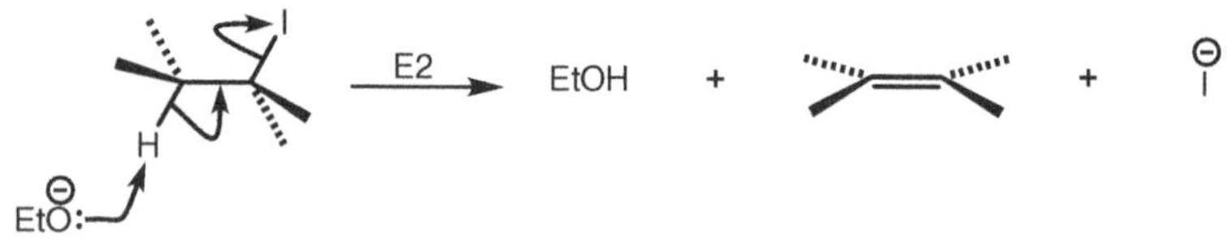

Equation 5

When checking a proposed mechanism, remember that the same atoms and overall charge must be present on *both* sides of the chemical equation; e.g. in equations 4 and 5, there is an overall charge of −1 on each side.

Orbitals add stereochemistry to curly arrows

Orbitals have their own spatial arrangements and their orientation *must* be carefully considered alongside a proposed curly arrow mechanism to take account of the orbital overlap needed in the reaction. The mechanism of nucleophilic displacement involving Me_3N and MeI is better written as:

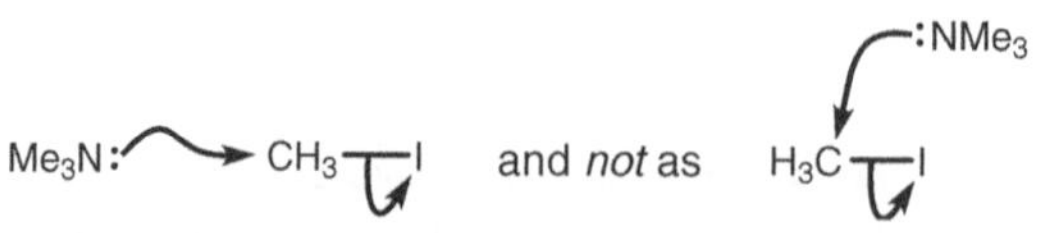

because orbital overlap is required between the non-bonded pair of electrons (HOMO) on N and the antibonding σ* orbital behind the C–I bond, otherwise the reaction will *not* occur.

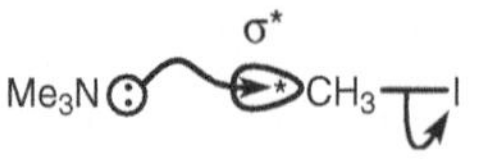

HOMO: N lone pair
LUMO: σ* of C–I

Nucleophiles, such as MeLi, approach a C=O bond from above (or below) the plane and to the rear of the C atom, where the lobe of the antibonding π* (LUMO) of the C=O bond is located:

and *not* like this 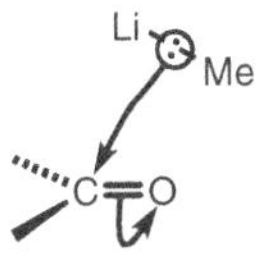   where there is no orbital overlap.

The ability of a non-bonded pair of electrons to conjugate with an adjacent π-system is controlled by the relative orientation of the two sets of orbitals. For example, in normal amides such as **A**, the non-bonded electrons on N are arranged so that they overlap and delocalise with the C=O π-bond. (This means that the N of an amide is much less basic than the N of an amine like **B** in which there is no delocalisation of the N non-bonded electron pair.) However, in the bicyclic amide **C**, the non-bonded pair on N cannot overlap and delocalise with the C=O π-bond. This explains why the N of amide **C** is much more basic than that of amide **A**, and is nearer in basicity to amine **D**.

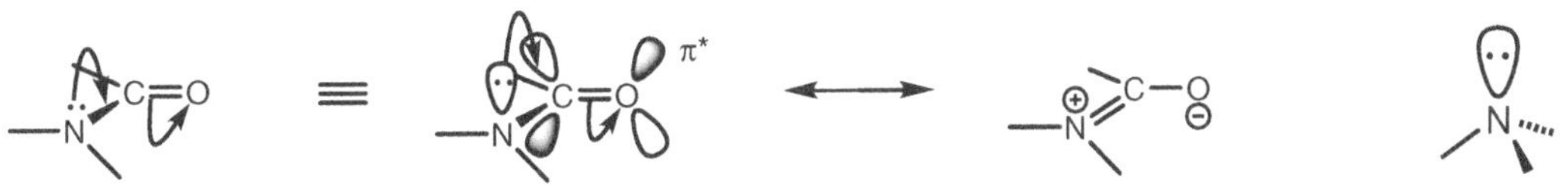

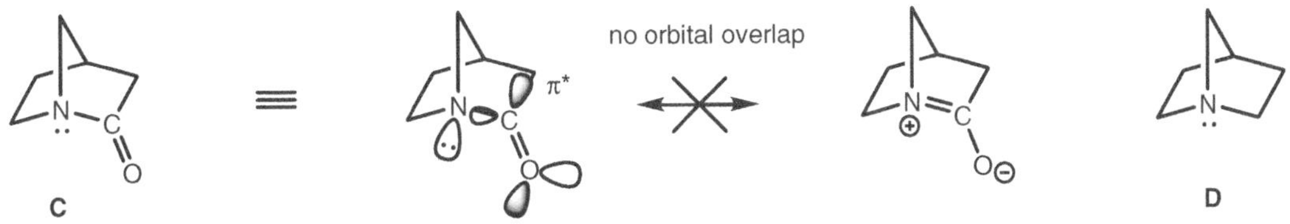

Electrophilic addition to a C=C bond occurs from above (or below) the plane, and in between both Cs where the electron density of the π-bond (HOMO) is highest. For example, electrophilic addition of H^+ to alkene **E** proceeds to give the carbocation intermediate **F**, where the incoming H and the resulting p-orbital have a coplanar relationship.

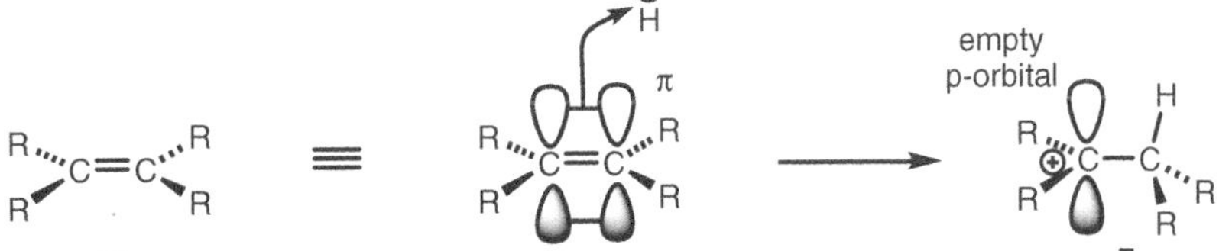

Drawing conventions‡

1. Acyclic structures

Line formulae such as $CH_3CH_2CH_2CH_3$ are used when we start organic chemistry, but these give no idea of three-dimensionality. We have been using perspective formulae so far (with ——— , ıııııı and ▬▬ bonds), but there are other conventions.

Remember that the hydrogen atoms are omitted in skeletal formulae, e.g. ∧∨∧ and ⟋⟍⟋ (double).

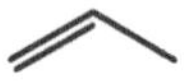

(a) Perspective formulae

Try to get the bond angles approximately correct; any convenient viewing direction will do.
For example:

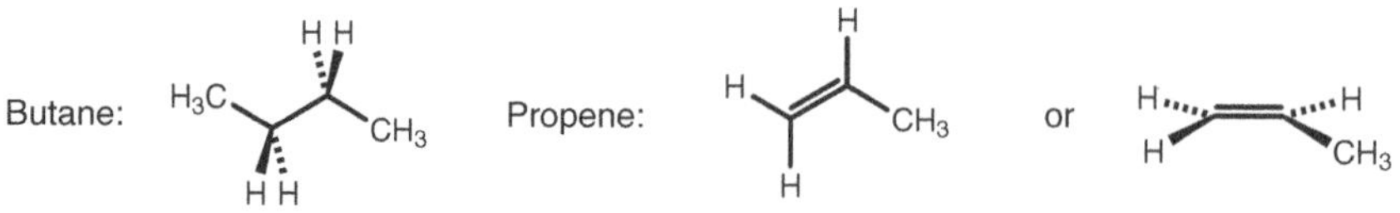

(b) 'Sawhorse' structures

These are used for saturated chains, especially for the two central carbons.
For example:

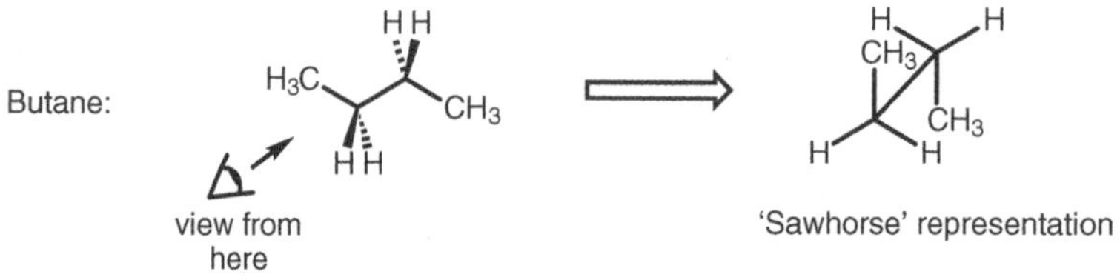

(c) Newman projections

Here the molecule is viewed along a carbon–carbon bond, which is shown as a circle. The bonds to the front carbon meet at the centre of the circle, those to the back carbon stop at the circumference.

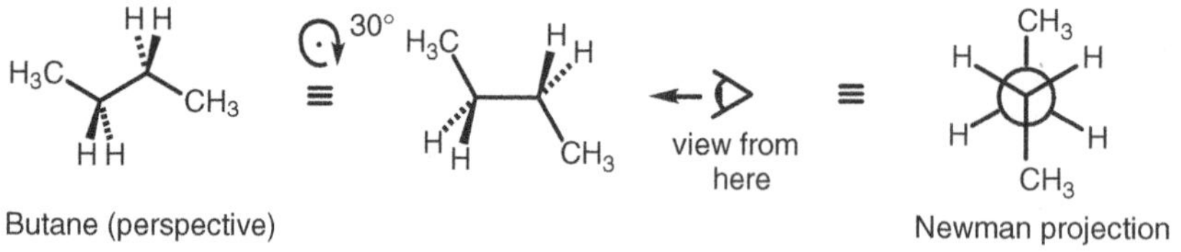

For double-bonded compounds, view along the C=C bond; the circle represents the double bond in this case.

These are good diagrams to show different conformations of open-chain structures and for reactions of alkenes because some arrangements, such as groups positioned anti- or synperiplanar to each other are especially easy to see. Single-carbon projections are good when assigning (*R/S*) descriptors (see page 16).

---

‡ For all symbols and abbreviations used, see inside back cover.

(d) Fischer projections

These are the most stylised and the least realistic, but they can often be useful. Each sp^3 carbon atom in the chain is written as a cross ┼, for which:

(i)   The main carbon chain must be in the *vertical* plane and pointing *away* from the observer.

(ii) The other two substituents must be in the *horizontal* plane and pointing *towards* the observer.

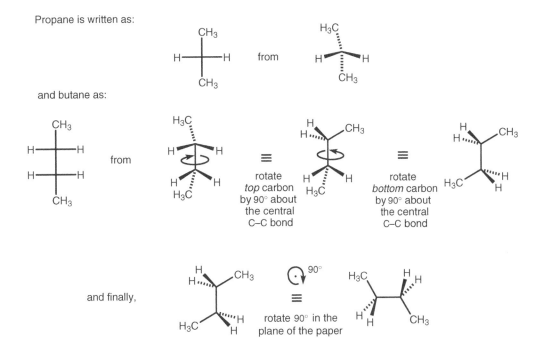

Fischer projections can be converted into Newman projections by viewing along the central C–C bond, either from top or bottom.

Notice that in a Fischer projection the correct bond angles (of 109°) are ignored. Do not be tempted to put a *C* atom at the centre of the cross, or it ceases to be a Fischer projection. These projections are good for quickly drawing structures of certain isomers, but they give little immediate idea of three-dimensionality.

The key points with these four conventions are:

- Learn how to use and interconvert drawings of *all* types, including linear structures.
- Use the type which (a) best suits you and (b) best suits the topic.

We will mostly use perspective drawings and Newman projections in this book.

2. Cyclic structures

Six-membered rings

How to draw a chair form of cyclohexane step-by-step

	Right handers may find this column easier	Left handers may find this column easier

(1) Draw two parallel sloping lines

then draw the 'nose' down at the 'top' end of the lines

and then the 'nose' up at the 'down' end. This is the carbon skeleton[1].

(2) Put in the *axial bonds* as vertical lines, up when the adjacent bonds point upwards

and down where the adjacent bonds point downwards.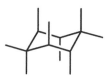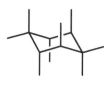

(3) Put in the *equatorial bonds* by completing the tetrahedra of bonds at each end

and completing the others with lines that point slightly down at the 'nose-down' end making a lying-down M shape

and up at the 'nose-up' end making a lying-down W shape

PRACTISE DRAWING THESE.

---

[1] This can be done continuously in one single line.

Newman projections of cyclohexane

These are taken by viewing along the two C–C bonds on opposite sides of the ring.

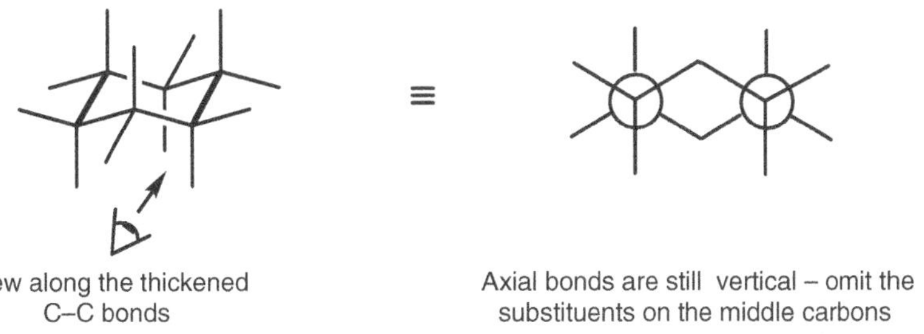

view along the thickened
C–C bonds

≡

Axial bonds are still vertical – omit the
substituents on the middle carbons

Check this using a molecular model kit.

Conformations of cyclohexane: chairs and boats

Cyclohexane is a flexible molecule. By rotation about its C–C bonds, it can convert itself from the structure in the right-hander's column to the structure in the left-hander's column on page 8. This can be seen using chlorocyclohexane as an example.

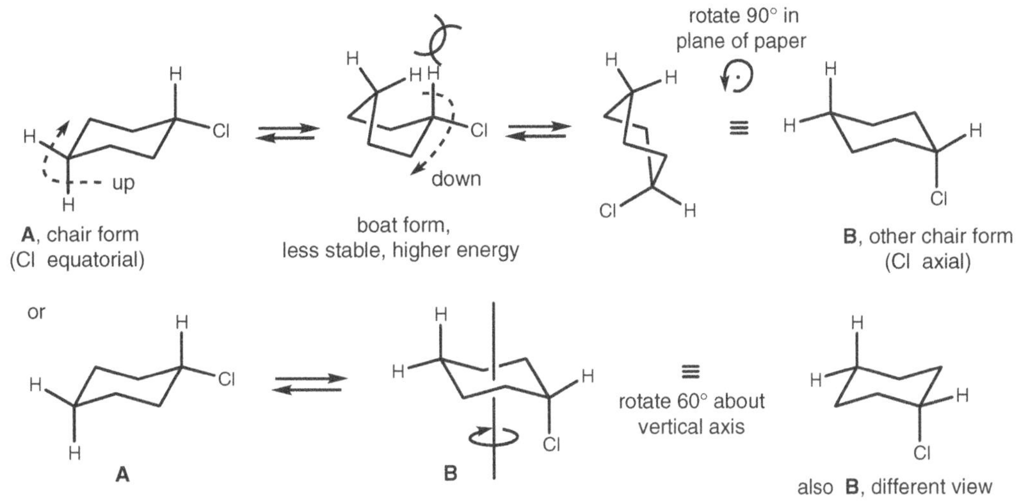

A, chair form
(Cl equatorial)

boat form,
less stable, higher energy

rotate 90° in
plane of paper

B, other chair form
(Cl axial)

or

A

B

rotate 60° about
vertical axis

also **B**, different view

This chair-to-chair conversion is called a *ring-flip* and you can draw it either as in the top line (if you are happy drawing both versions of the chair) or as in the bottom line (in which the substituents look as though they have all moved one position around the ring; this is useful if you draw one chair better than the other). Check it by making molecular models!

The intermediate boat form is normally much less stable than the chairs. The *twist-boat*, formed when the two end carbons of the boat are twisted away from each other to minimise the close proximity of the substituents across the ring, is slightly more stable than the symmetrical boat.

A single-carbon bridge across the boat gives the strained, rigid [2.2.1]-bicycloheptane (norbornane) system.

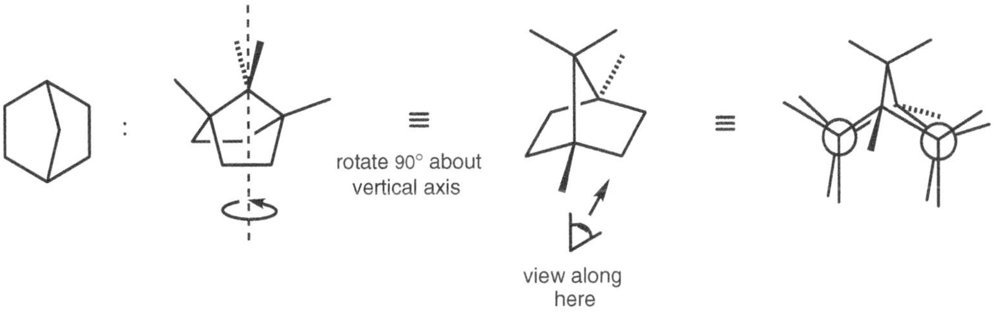

rotate 90° about
vertical axis

≡

view along
here

≡

[2.2.2] structures can be drawn in a similar way using a boat conformer:

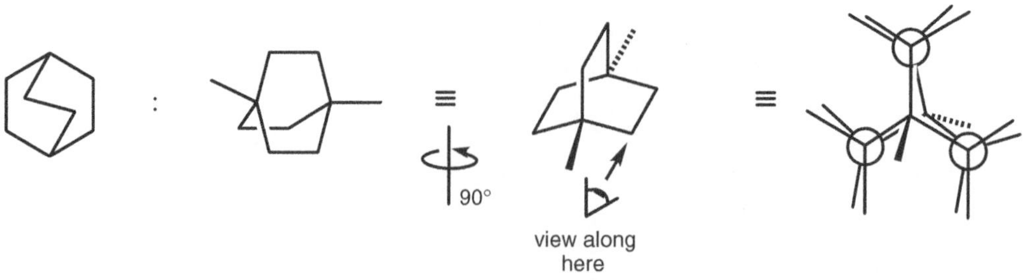

If you can draw cyclohexane, you can draw almost any ring structure in organic chemistry!

Fused cyclohexanes: the decalins

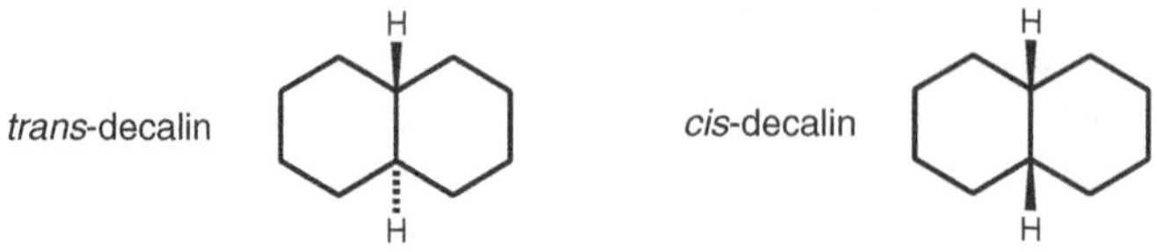

trans-decalin

cis-decalin

trans-Decalins:

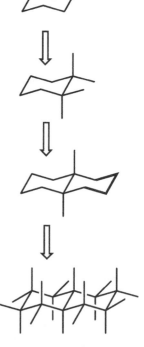

(1) Draw a chair cyclohexane.

(2) Put in the axial and equatorial bonds on adjacent carbons.

(3) Complete another chair using the two equatorial bonds (the axials will carry the bridge substituents).

(4) Fill in the axial and equatorial bonds as for cyclohexane.

trans-decalin is a rigid, 'flat-ish' structure

Compounds with more *trans*-fused rings can be drawn in a similar way, e.g.

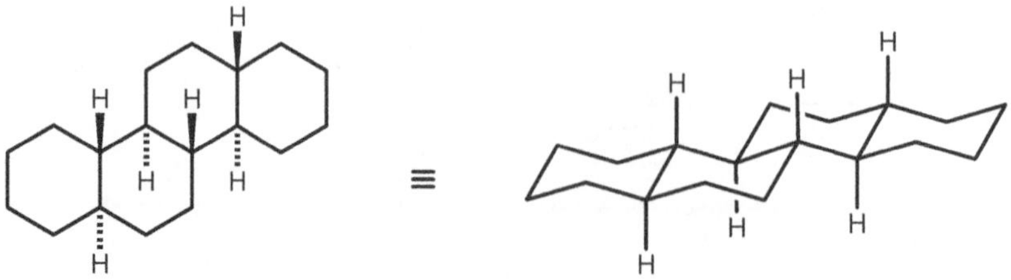

cis-Decalins:

This is a flexible structure with two major enantiomeric conformations.

(1) *First conformer*
     Draw the cyclohexane skeleton.

     Put in both substituent bonds on adjacent carbon atoms.

     Complete another chair using the downward-pointing axial and equatorial
     bonds.

     Finally, put in the substituents as usual for cyclohexane rings. When you do the left-hand ring, rotate your paper
     so that this ring looks like an ordinary cyclohexane.

(2) *Second conformer*
     This time draw this cyclohexane

     Then complete using the downward-pointing axial and equatorial bonds, etc.

The two major enantiomeric conformers of *cis*-decalin can interconvert via a double boat form.

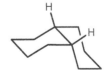

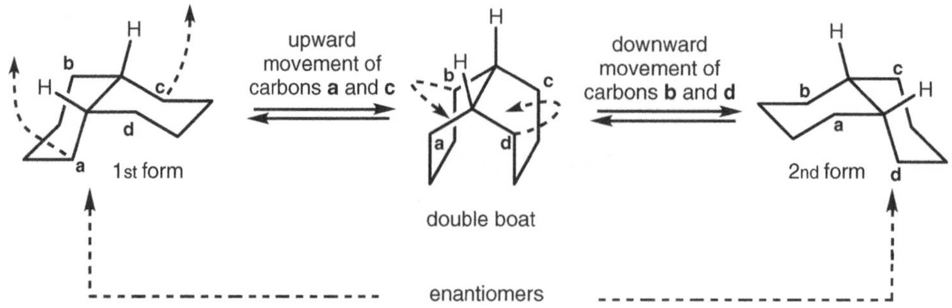

double boat

1st form

2nd form

enantiomers

Try this using molecular models.

These double-chair structures can also be drawn using the letter Z (and its mirror image) as the central bridge.

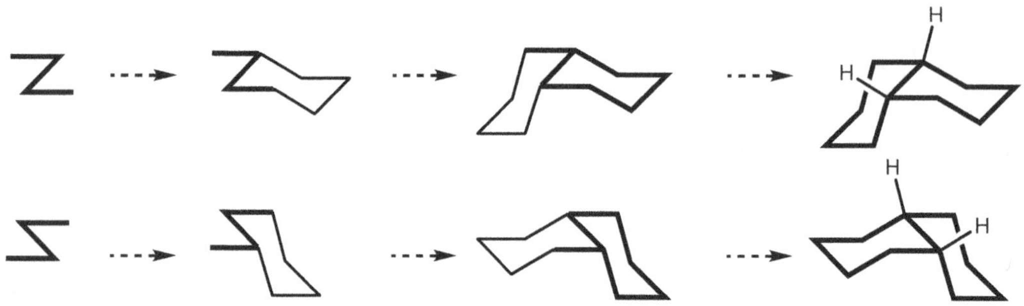

Cyclohexenes

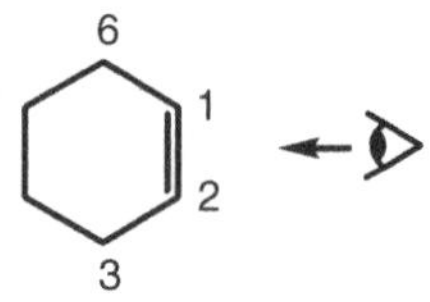

In cyclohexene, atoms 1, 2, 3 and 6 are co-planar. The ring is part planar, part puckered. To represent this molecule viewed along the alkene plane: first draw the four carbon atoms 3, 2, 1 and 6 (with the double bond) in a line.

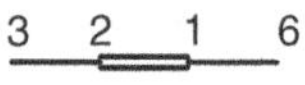

then put C(4) as a dot above C(2) and put C(5) as a dot below C(1); (or vice versa).

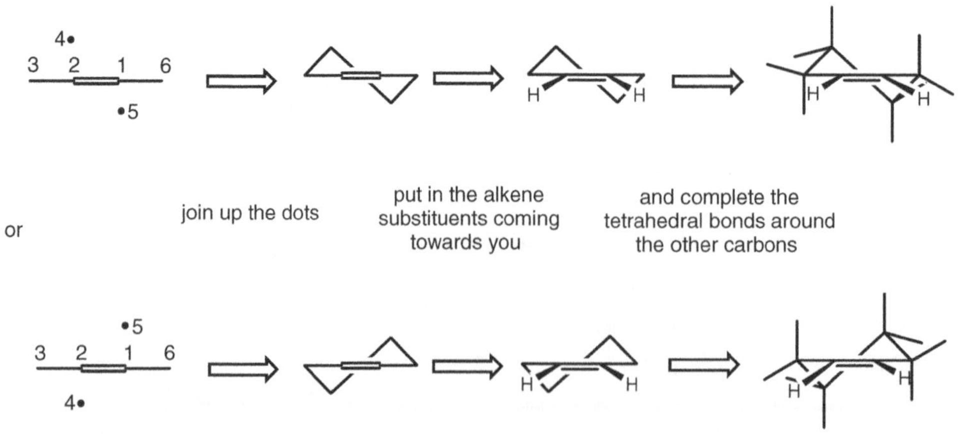

or

join up the dots

put in the alkene substituents coming towards you

and complete the tetrahedral bonds around the other carbons

These are the two major half-chair enantiomeric conformers of cyclohexene.

Three, four and five-membered rings

These can be drawn by chopping off the correct-sized piece of cyclohexane.

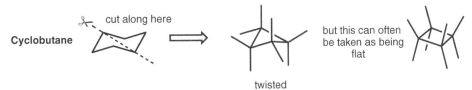

This type of cut also gives a cyclopentane framework – see below

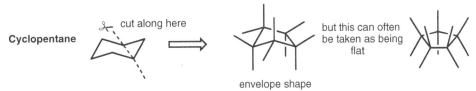

This type of cut also gives another cyclobutane framework

This type of cut also gives a cyclopropane framework – see above

Seven- and eight-membered rings, and larger rings

Continue the two *zig-zag* parallel lines of carbon atoms from cyclohexane until there are enough atoms. Finally fill in the substituents (as for cyclohexane).

Cycloheptane delete the non-bold bond of cyclohexane **1** to give **2**, add one more bond to carbon to give **3**, and join up the ring in **3** to give cycloheptane **4**.

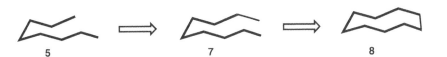

Cyclooctane repeat the procedure above, starting from **3**.

Cyclononane repeat the procedure above, starting from **5** but add the bond to the other side to give **7**. And so on to give **8**.

Remember that these rings are quite flexible compared with cyclohexane (use molecular models to check this).

Conformational isomers

So far this chapter has dealt with the drawing of major conformers associated with cycloalkanes. These conformations are much more predictable than for related acyclic (open chain) alkanes, due to the presence of a carbocyclic ring which reduces the overall number of possible conformational isomers.

Conformational isomers are isomers which are easily interconverted by a low energy process, often by rotation about single bonds at room temperature (25°C). These easily varied spatial arrangements of atoms in a molecule can give rise to an infinite number of conformational isomers. Some of the more important ones are described below.

In each example, view the molecule as a Newman projection along a C–C bond and look at the dihedral angles made between bonds on adjacent carbon atoms.

Ethane

Staggered   Dihedral angle 60° between any two bonds.

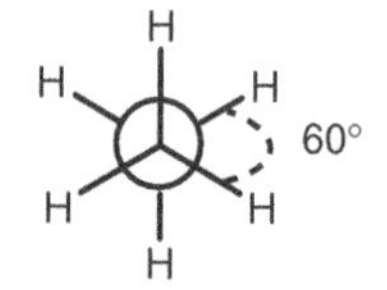

Eclipsed   Dihedral angle 0° between any two bonds.

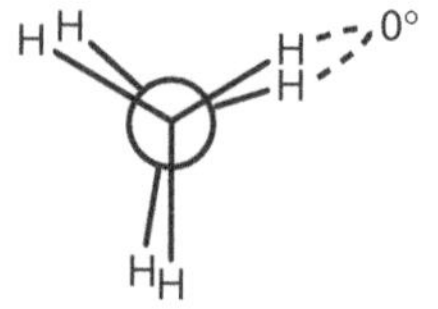

Butane

Antiperiplanar   A staggered conformation, dihedral angle 180° between two (named/large) substituents.
e.g. the methyl groups in butane.

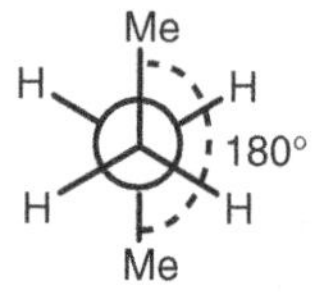

Gauche   A staggered conformation, dihedral angle 60° between two substituents.

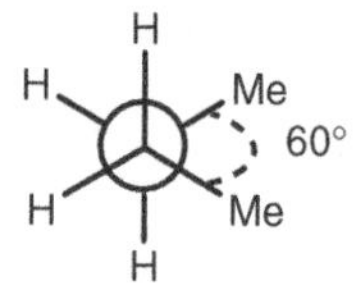

Synperiplanar   An eclipsed conformation, dihedral angle 0° between two substituents.

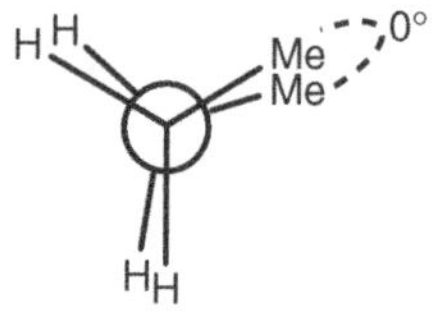

Atropisomers: chirality due to restricted rotation

When rotation about single bonds is seriously restricted by the size of nearby groups, different conformers (atropisomers) can be isolated. *Ortho*-substituted biphenyls in which each ring bears two different *ortho*-substituents can exist as separable enantiomers. The rate of racemisation depends on the size of the groups and the temperature. For example:

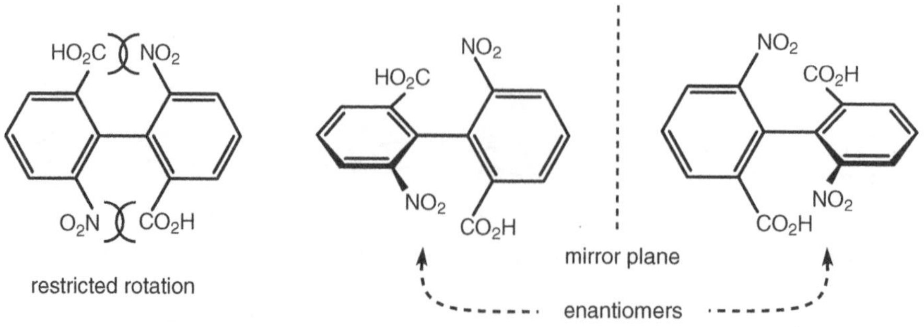

restricted rotation   mirror plane   enantiomers

Stereochemical terms

ACHIRAL: an object which is *superimposable on its mirror image*.
For example:

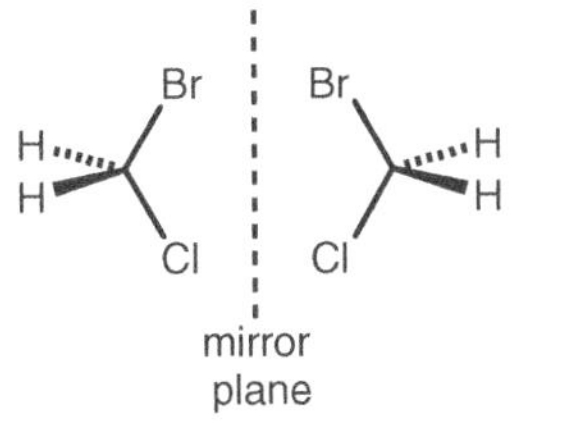

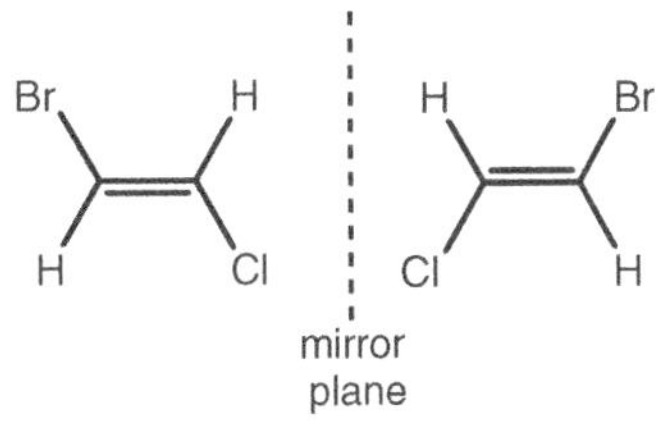

None of these molecules will rotate the plane of plane polarised light.

CHIRAL: an object which is *non-superimposable on its mirror image*.
For example:

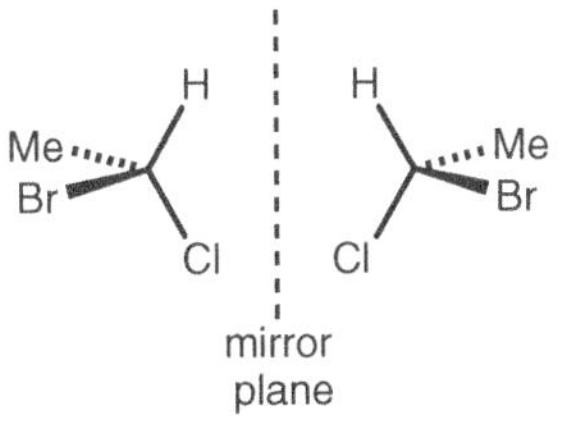

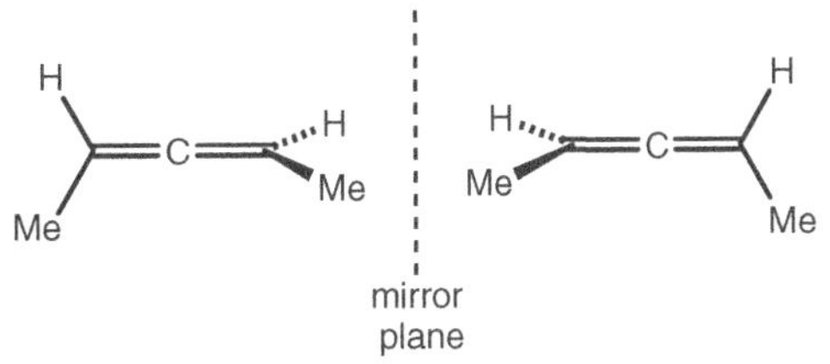

In each pair, the two molecules will rotate the plane of plane polarised light equally but in opposite directions. A chiral object cannot have a plane or centre of symmetry. Chiral molecules rotate the plane of plane polarised light unless both mirror images are present in equal ratio (racemic mixture).

CHIRAL CENTRE (stereogenic centre or chirality centre): that part of a structure which gives rise to chirality. It is often a carbon atom substituted with four different groups, *but not always*; see quaternary ammonium ions and allenes. Molecules that contain chiral centres may or may not be chiral overall (see isomers on page 19).

ENANTIOMER: an object which has a non-superimposable mirror image (see chiral, above).

PLANE OF SYMMETRY or MIRROR PLANE: an imaginary plane that bisects an object into a pair of mirror image fragments. For example:

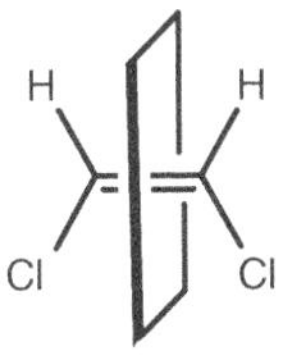

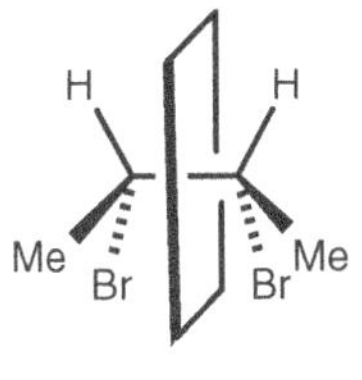

CENTRE OF SYMMETRY or CENTRE OF INVERSION: if you take an imaginary line from any part of an object through this point and on beyond an equal distance, you end at an identical part of the object; denoted by a large dot. For example:

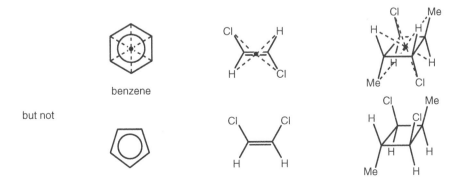

benzene

but not

ROTATION AXIS: Two-Fold Rotation Axis (C_2) – this is an axis about which a rotation of 360°/2 (=180°) will arrive at an identical fragment. For example:

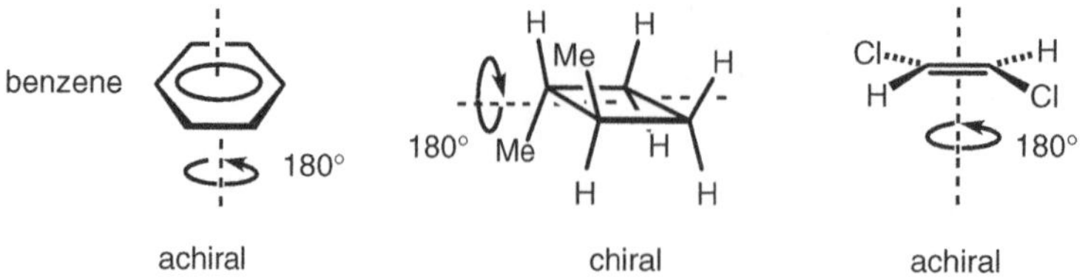

STEREOCHEMICAL DESCRIPTORS: Cahn-Ingold-Prelog (CIP) system. (*R*)-/(*S*)- are used for chiral centres and (*E*)-/(*Z*)- are used for geometric isomers. In this system, substituents are first given priorities.

(1) Look at the atomic number of the first atom attached;
The higher the atomic number the higher the priority: e.g. I > P > O > H.
Also, the heavier the isotope the higher the priority: T(^{3}H) > D(^{2}H) > H.

(2) If all substituents are not distinguished using the rules outlined in the first step, then proceed to the next set of atoms.
 (a) One high-priority atom takes precedence over several low; e.g. $CH_2SH > CPh_3$.
 (b) Multiple bonds to an atom are counted as the equivalent number of single bonds;
 e.g. a C=O is equal to two C–O bonds; so $CHO > CH_2OH$;
 and a C≡C is equal to three C–C bonds; so $C{\equiv}CH > CH_2CH_3$.

It is important to note that a group's priority is not determined by the sum of its atomic numbers or by its size!

(*E*)-/(*Z*)- (for geometric isomers): If the higher priorities are on the same side of the double bond then it is denoted as (*Z*)-; if they are on opposite sides, this is denoted as (*E*)-.

⟹

Position	C(1)	C(2)
High priority	**Br**	**Cl**
Low priority	F	H

⟹ so this has a (*Z*)-configuration

⟹

Position	at C(1)		at C(2)	
Group	Ph	CH_3	CHO	CH_2SH
First atom	**C**	**C**	**C**	**C**
Second atom	**C**,C,C	**H**,H,H	**O**,O,H	**S**,H,H
Priority order	High	Low	Low	High

⟹ and this has an (*E*)-configuration

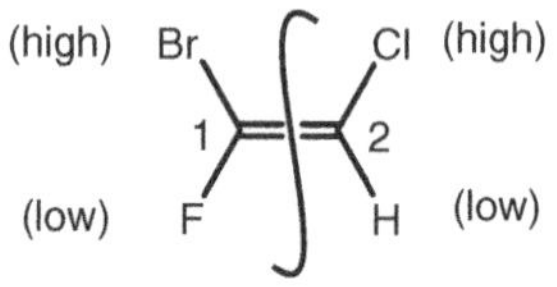

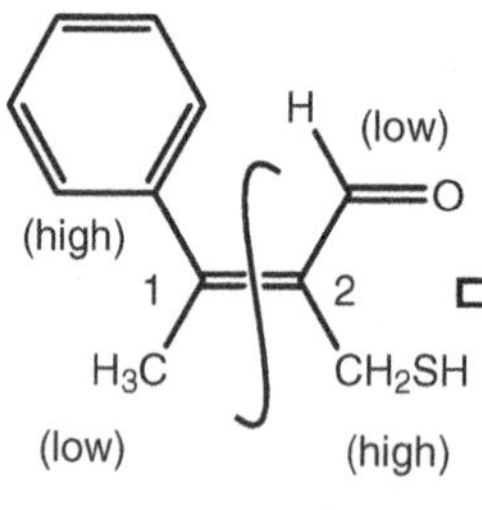

(R)-/(*S*)- (for chiral centres): First, assign priorities to the four substituents attached to the chiral centre, as before. Next arrange to view the chiral centre with the lowest priority substituent at the *back* of the centre (usually the atom) concerned. If the remaining substituents viewed are in descending priority order clockwise, it is an (*R*)-centre; if they are anticlockwise, it is (*S*)-.

For example:

Four groups	CH₃	CO₂H	NH₂	H
First atom	C	C	**N**	**H**
Second atom	**H,H,H**	**O,O,O**		
Priorities	3	2	1	4

H (lowest priority) at *back* of Newman projection

Clockwise, so (*R*)-

CONFIGURATIONAL ISOMERS: isomers which are *not* interconverted by a low energy process at room temperature. These are typically stereoisomers which contain chiral centres or centres of geometrical isomerism due to the presence of a double bond or ring. For interconversion to occur, cleavage of a σ- or π-bond is required. These spatial arrangements of atoms in a molecule give rise to a fixed number of configurational isomers; the theoretical maximum being 2^n, where n is the number of fixed configurational centres within the molecule. However, the number of isomers may be lower than this due to the presence of a plane or point of symmetry, or a ring within a given stereoisomer. It is interesting to note that a molecule with a fixed configuration can have an infinite number of conformations, whereas a molecule of fixed conformation can *only* have a fixed number of configurations.

RACEMIC compound: an equimolar mixture of two enantiomers – this is often termed a racemate. See also the note after relative configurations on page 19.

RACEMISATION: the process of converting one enantiomer into a 1:1 mixture of that enantiomer and its non-superimposable mirror image isomer (a racemic mixture).

For example:

(*R*)-

1:1 ratio, (*R*)-:(*S*)-

RESOLUTION: the process of separating a racemic mixture into its two enantiomers.

OPTICAL PURITY: a term used to describe the excess of one enantiomer in a solution containing two enantiomers in unequal amounts. This can simply be calculated by subtracting the percentage of the minor enantiomer from the major enantiomer. The excess can also be described as the ENANTIOMERIC EXCESS.

For example:

100:0 ratio, (*R*)-:(*S*)-

(*R*)-
optically pure

60:40 ratio, (*R*)-:(*S*)-

(*R*)-
20% optical purity

SCALEMIC: a term used to describe an optically impure mixture due to the presence of an unequal proportion of both enantiomers. For example:

75:25 ratio, (*R*)-:(*S*)-

a scalemic mixture containing the
(*R*)-enantiomer: optical purity = 50%

OPTICAL ROTATION: if an optically active compound rotates the plane of plane-polarised light in a clockwise direction [(+)-rotation], it is dextrorotatory. Whereas, its enantiomeric form will be laevorotatory, allowing the rotation of the plane of plane-polarised light in an equal amount but in an anticlockwise direction [(−)-rotation]. It is important to note that the sign of rotation (+)- or (−)- is independent of the configuration of the chiral centre(s) present. An achiral, *meso*- or racemic compound will NOT rotate the plane of plane-polarised light – they are optically inactive.

DIASTEREOISOMER: a stereoisomer that has a non-superimposable non-mirror image.

MESO-COMPOUNDS: these are *achiral but contain at least two chiral centres* and usually have a plane or centre of symmetry, such as (*S,R*)-MeCH(Br)CH(Br)Me. In all *meso*-compounds, each type of (*S*)-centre is matched by an (*R*)-centre. The *meso*-compound does *not* rotate the plane of plane-polarised light.

For example:

EPIMER: these have two or more chiral centres but differ in the stereochemistry about only *one* of these chiral centres, such as these two triols **A** and **B**.

These Fischer projections can also be drawn as perspective diagrams, but the process of conversion is quite lengthy. For example:

Then convert this to the standard *zig-zag* carbon chain

RELATIVE CONFIGURATION: the configuration at any chiral centre with respect to that at any other chiral centre in the molecule (or in a closely related molecule). For example, see the beginning of question (7) on page 56, where the deduction from the chemistry is that the *relative* configuration of **B** is the same as **A**, despite the *absolute* stereochemistry of **A** being as yet unknown.

Note: When a pair of enantiomers undergoes the same stereospecific reaction, the change in relative stereochemistry is identical for each enantiomer. For this reason (and to save space) organic chemists often draw only *one* enantiomer to represent the behaviour of the pair. Where relevant in this book, racemic compounds will be labelled *racemic* or given the prefix (±); single enantiomers will be labelled as such or given their (*R/S*) prefix.

ISOMERS

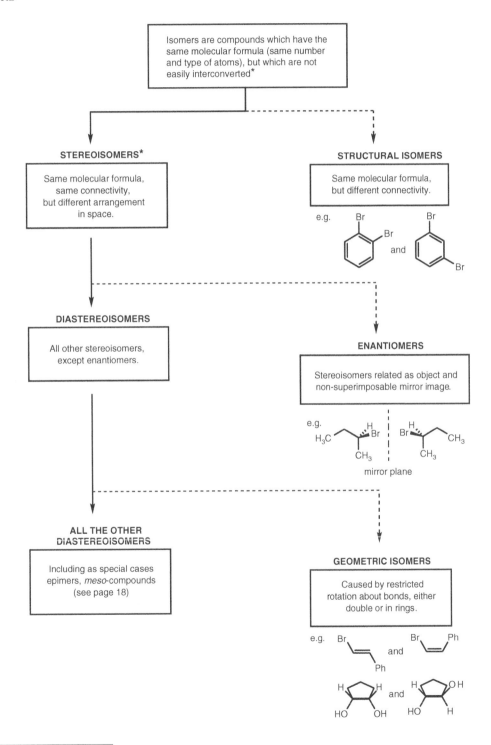

---

*See page 14 for conformational isomers, which *are* easily interconverted at room temperature.

Selectivity

A *selective* synthesis gives a higher yield of one (or more) of a number of possible products.
Regioselectivity or *positional selectivity* is selectivity between possible *structurally isomeric* products.
For example:

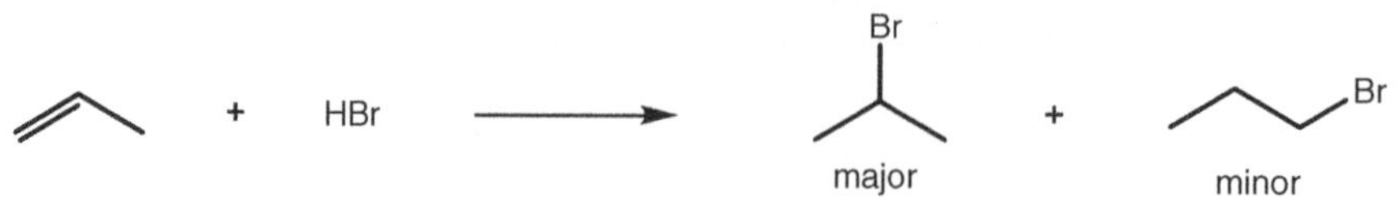

(think about Markovnikov's mnemonic as an example)

Stereoselectivity is selectivity between possible *stereoisomeric* products.
For example:

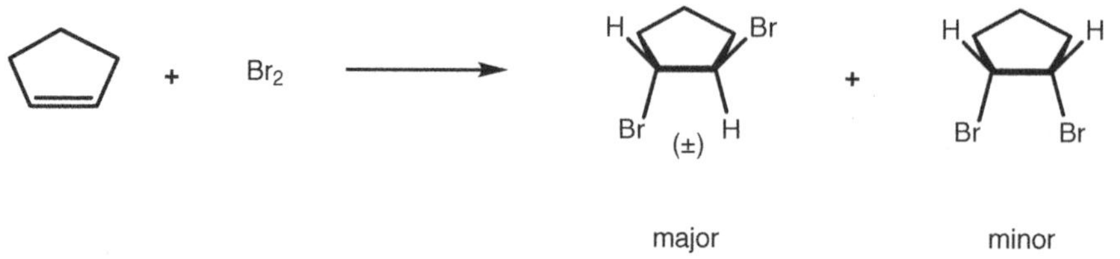

In a *stereospecific* synthesis, stereoisomeric starting materials **A** and **B** give stereoisomerically different products under the same conditions. That is, stereoisomer **A** forms stereoisomer **X** but not **Y**, and stereoisomer **B** forms stereoisomer **Y** but not **X**. A more comprehensive discussion is given on page 21.

Enantioselectivity is selectivity between possible *enantiomeric* products.
For example:

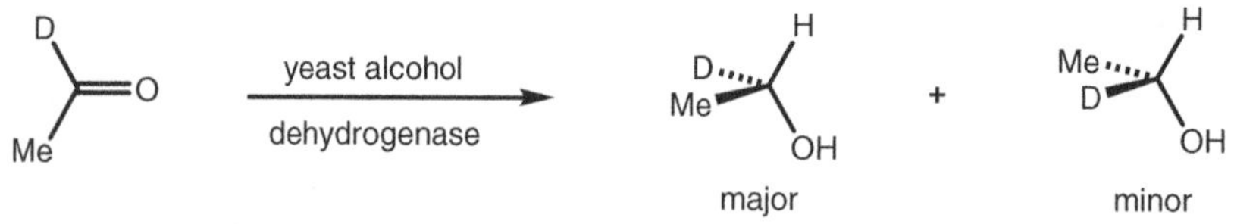

Diastereoselectivity is selectivity between possible *diastereoisomeric* products.
For example:

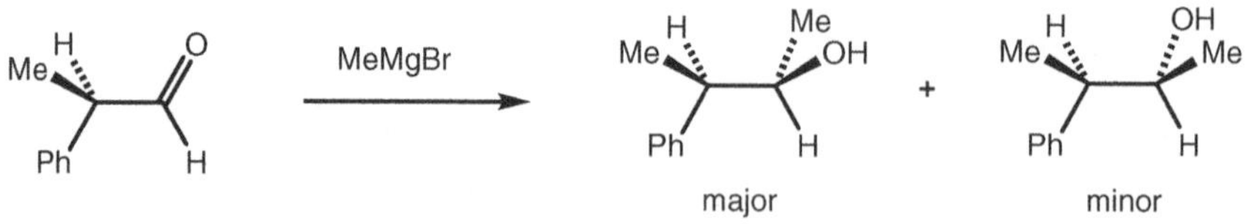

Chemoselectivity is selectivity between similar *functional groups*.
For example:

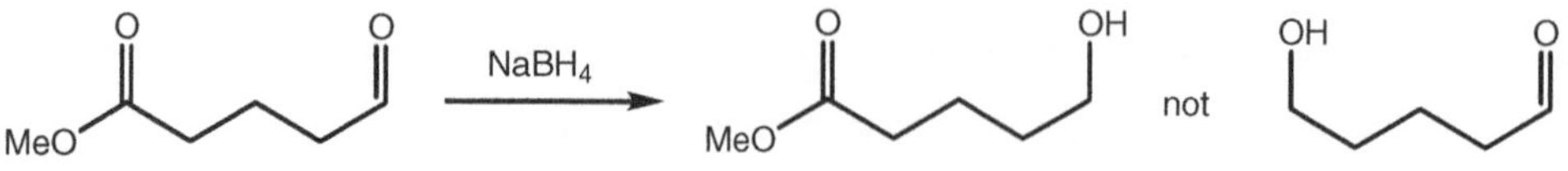

Stereospecificity versus stereoselectivity

To determine whether a given reaction is stereoselective or stereospecific can sometimes be rather tricky. If a particular reaction gives a mixture of stereoisomeric products, is this process automatically stereoselective? Or, if only one stereoisomeric product was isolated, does this process have to be stereospecific? In order to answer these questions, we need to look at the definitions of stereoselectivity and stereospecificity.

The terms stereoselectivity and stereospecificity have generally been used to describe the stereochemistry and product composition of a particular reaction. However, these stereochemical descriptors lend themselves much better to describing the stereochemical outcome of a given mechanism than the product distribution. Therefore, the easiest way of deducing whether a given reaction is stereospecific or stereoselective, is to look at the mechanism and *not* the product(s).

If there is a choice in the stereochemical pathway, whether that be 5 to 1, or 1000 to 1, the reaction is *stereoselective*, if there is *no* choice then the reaction is simply *stereospecific*.

For example,
(a) Azide displacement of chloride (R)-**A** gave exclusively azide (S)-**B**. There is *no* choice in the reaction pathway; the reaction *has* to proceed via S_N2 displacement leading to inversion of configuration as shown in **C**. Therefore, the reaction is *stereospecific*.

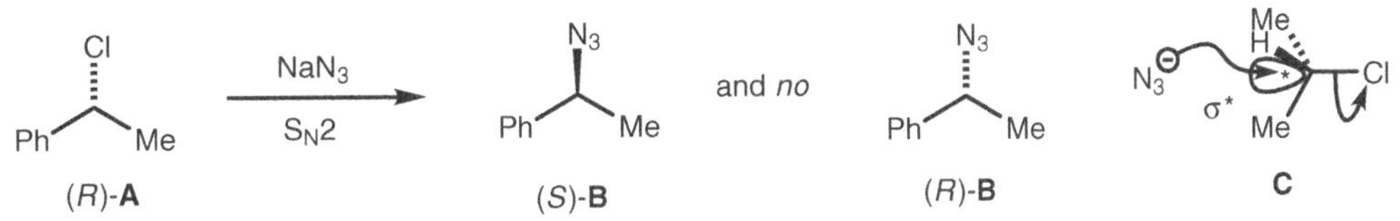

(b) Sodium borohydride reduction of ketone **D** gave a diastereoisomeric mixture of alcohols *anti-* and *syn-***E** (ratio 90:10). There is clearly a choice in the reaction pathway; addition can either occur on the top or bottom face of the carbonyl group in **F** to give *anti-* and *syn-***E** respectively. Addition to the top face is preferred (see pages 98 and 100), so this reaction is highly *stereoselective*.

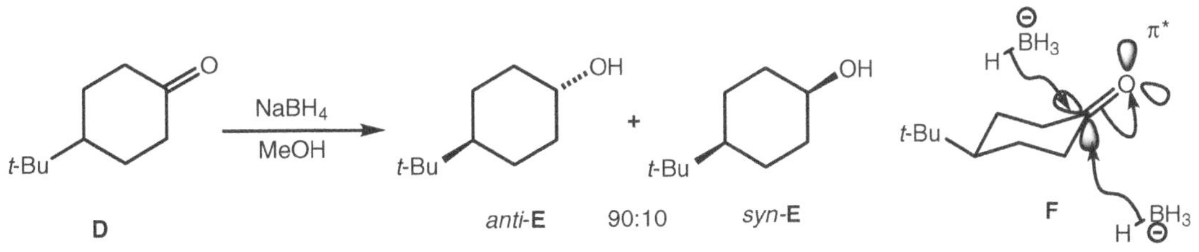

(c) Acetate displacement of tosylate (R)-**G** gave a racemic mixture of **H**. There is a choice in the reaction pathway; addition can either occur from the bottom face of the carbocation **I** (to give the (S)-enantiomer of **H**) or the top face to give the complementary (R)-enantiomer. Because both enantiomers are favoured, this reaction is *not stereoselective* since neither enantiomer is preferred.

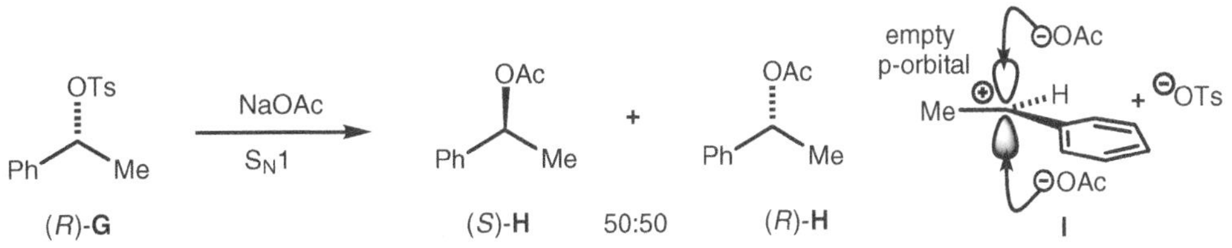

For reactions that appear to give a single stereoisomer (>99%) these could actually be stereospecific (and only one stereoisomer present) or very highly stereoselective, where the amount of the minor stereoisomer is very difficult to determine (due to limits of detection). The only way of determining whether these reactions are stereospecific or stereoselective is to determine whether there is a stereochemical choice within the reaction mechanism.

Problems for Chapter 1

1. Draw the following linear structures in a perspective projection.
   (a) $(CH_3)_2CHCH_2CH_2CO_2H$ (b) $CH_3OCH_2C(CH_3)_3$ (c) $HSCH_2CH_2CH(NH_2)CO_2CH_3$

2. Convert the following perspective projections of 2,3-dihydroxybutane into Newman projections viewed along the C_2–C_3 bond and comment on their relative energies.

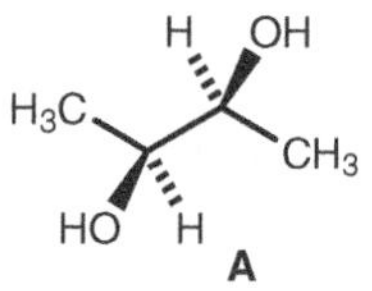

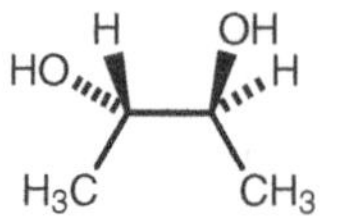

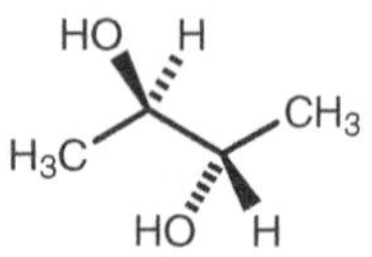

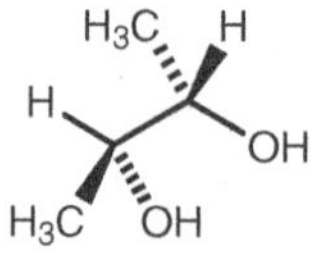

3. Re-draw the following linear structures as both perspective and Newman projections, illustrating the most stable conformation.
   (a) Cl_3CCH_3 (b) $PhCH_2CH_2CH_3$ (c) $(CH_3)_2CHCH_2OH$

4. Draw diagrams to illustrate the following:
   (a) *meso*-2,3-Dichlorobutane with an H and a Cl antiperiplanar.
   (b) (*R,R*)-2,3-Dihydroxybutane with the two OH groups synperiplanar.
   (c) (*R*)-2-Hydroxypropanoic acid with the two OH groups synperiplanar.

5. Re-draw the following Newman projections as perspective projections.

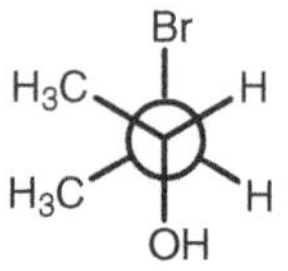

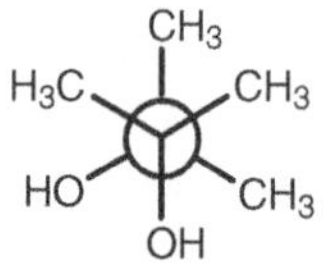

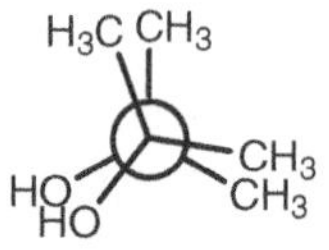

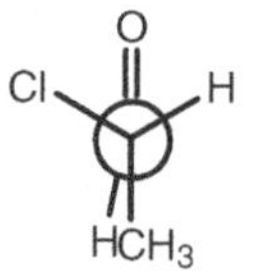

6. From the following compounds, pick out:
   (a) two with a plane of symmetry;
   (b) two with a centre of symmetry;
   (c) one that is chiral.

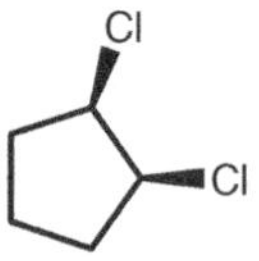

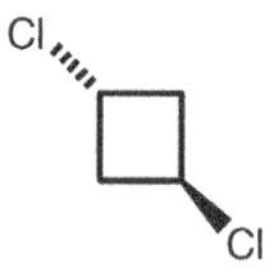

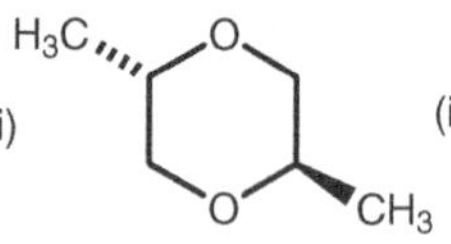

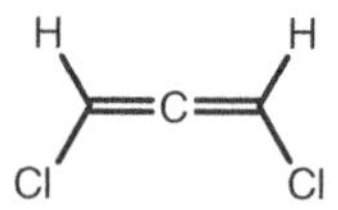

7. State the stereochemical relationship of the configurational isomers represented by **J**, **K** and **L** to compound **I**. (It may help to assign (*R/S*)- descriptors to each chiral centre.)

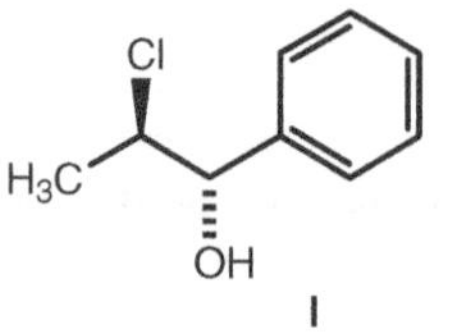

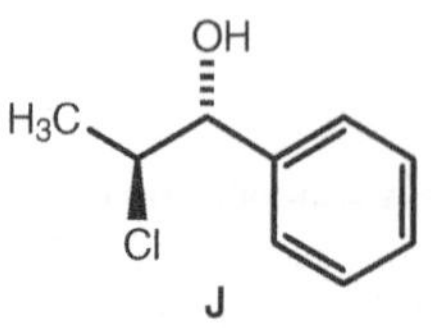

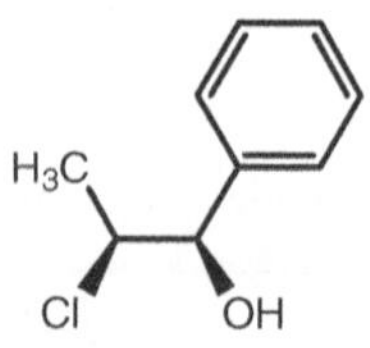

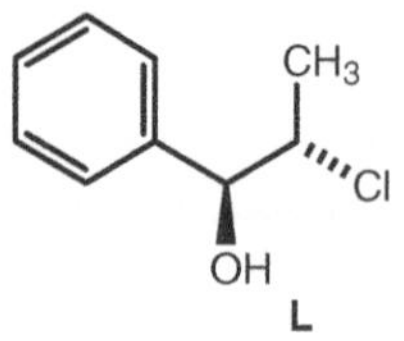

8. State the stereochemical relationships in the following groups of isomers.

(a)

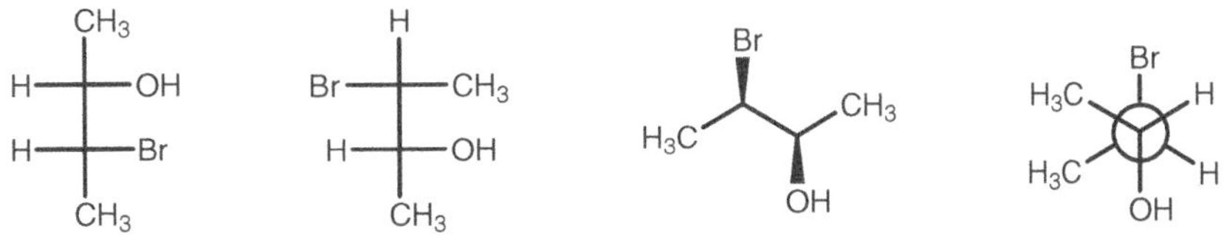

(b)

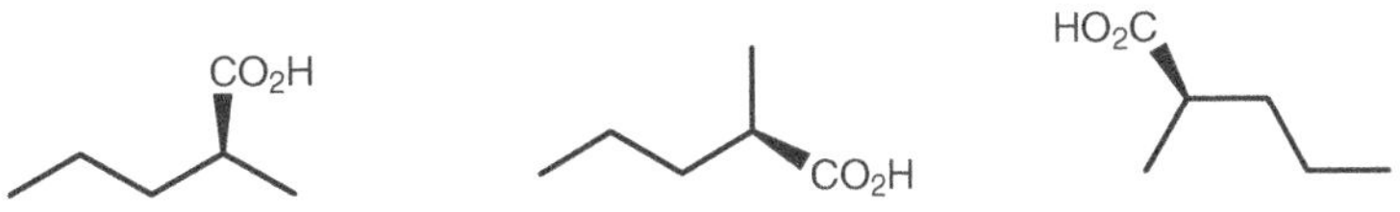

(c)

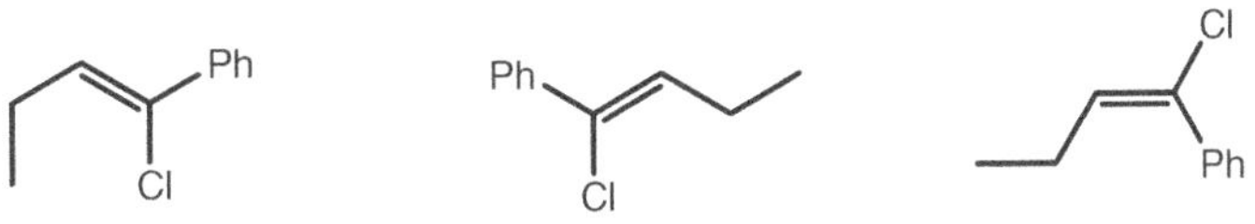

(d)

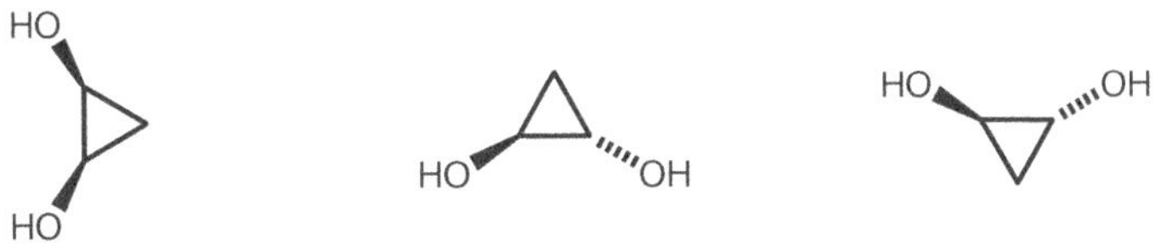

9. From the following molecules, pick out:
    (a) two which are achiral and contain no chiral centres;
    (b) two which are achiral and contain chiral centres;
    (c) two which are chiral and contain chiral centres;
    (d) two which are chiral and contain no chiral centres.

(i)

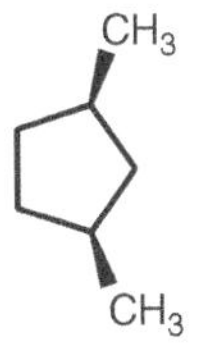

(ii)

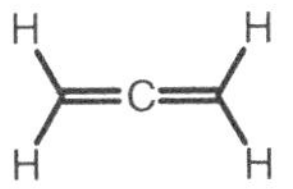

(iii)

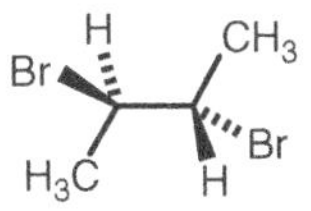

(iv)

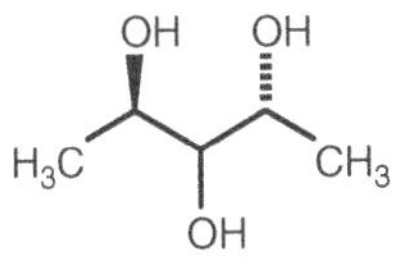

(v) 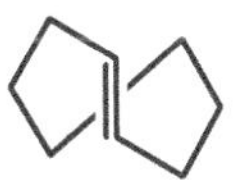 (*trans*-cyclooctene)

(vi)

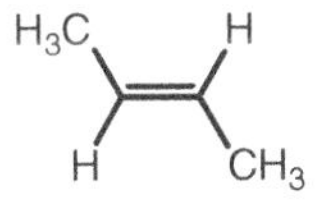

(vii)

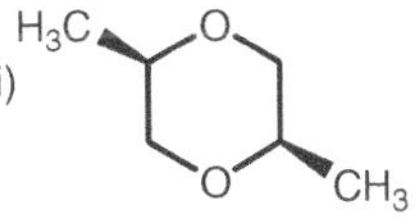

(viii)

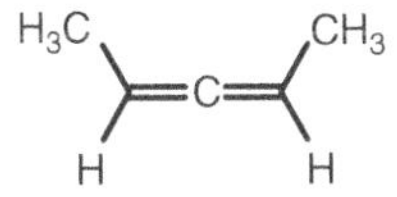

10. Draw clear diagrams to show all stereoisomers (other than conformations) of the following compounds, and state the stereochemical relationships between these isomers.

(a)

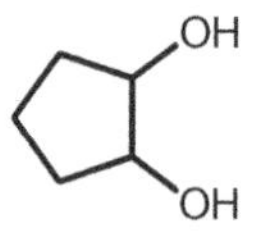

(b)

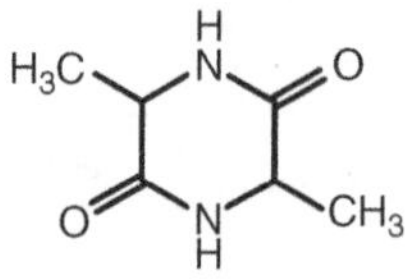

(c) $CH_3CH=C=CHCH_3$

(d)

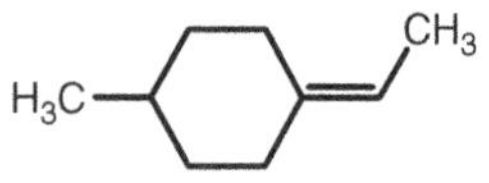

(e)

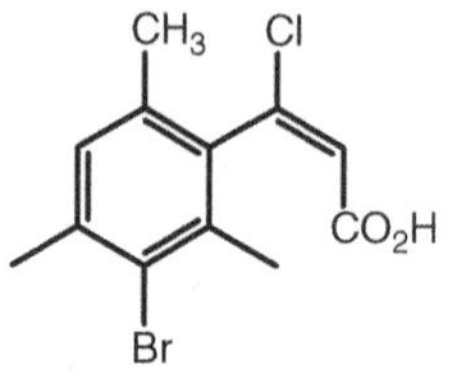

(f)

(g)

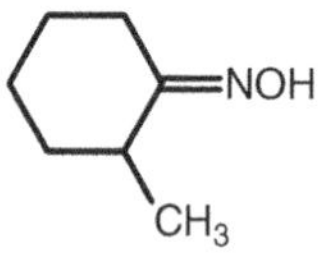

(h)

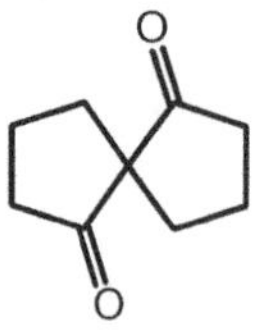

(i) 2-bromo-*trans*-decalin

(j) 2,2′ di-*tert*-butyl biphenyl

11. For the following molecules, predict:
    (a) the number of possible stereoisomers;
    (b) the number of possible pairs of enantiomers;
    (c) the number of possible diastereoisomers;
    (d) the number of possible epimers.

(i)

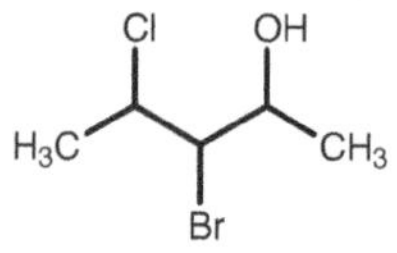

(ii)

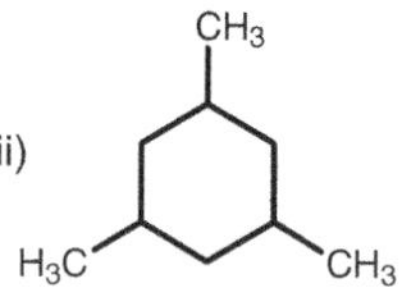

(iii)

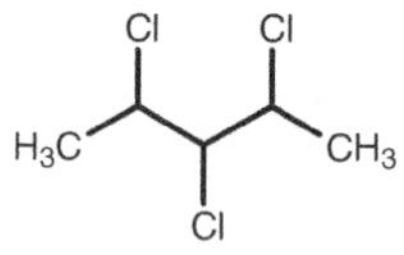

Answers for Chapter 1

1. First convert the linear structure into skeletal structure, then add the necessary perspective at each carbon atom. If you have a choice of whether to position the substituent *coming towards you* (on a wedged line) or *going away from you* (on a dashed line) then draw both forms.

(a) $(CH_3)_2CHCH_2CH_2CO_2H$

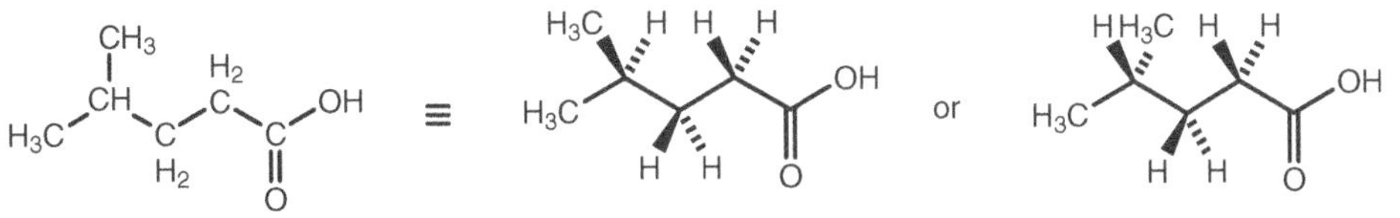

(b) $CH_3OCH_2C(CH_3)_3$

(c) $HSCH_2CH_2CH(NH_2)CO_2CH_3$

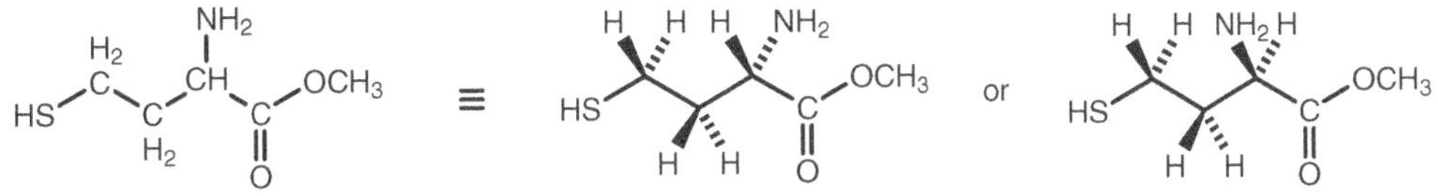

2.

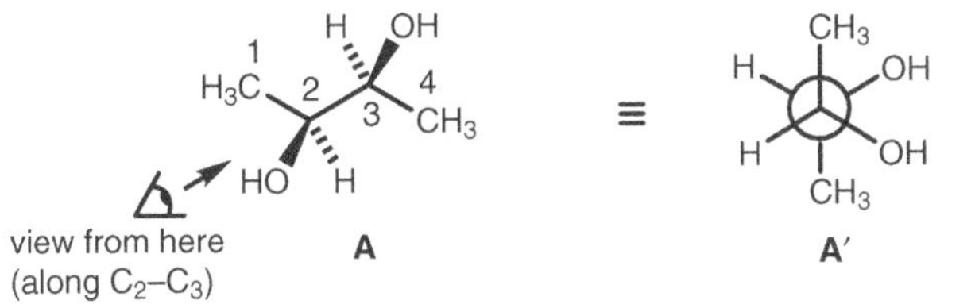

There are three gauche interactions present in this conformer; $2 \times CH_3/OH$ and OH/OH.

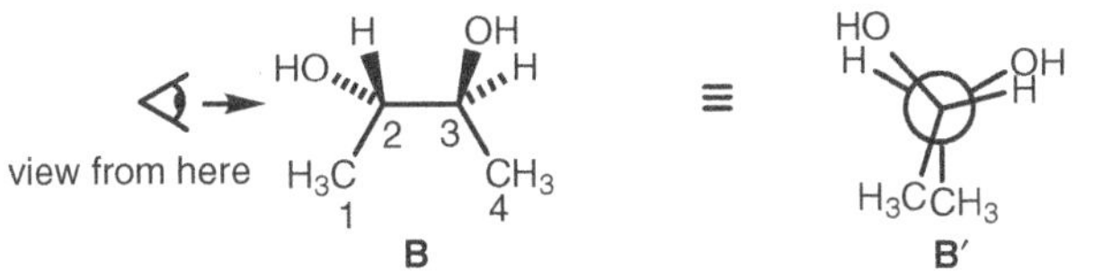

This is a high energy eclipsed conformer.

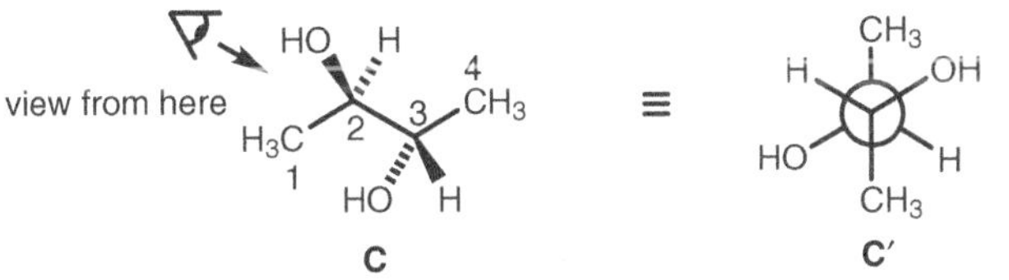

There are two gauche interactions present in this conformer; $2 \times CH_3/OH$.

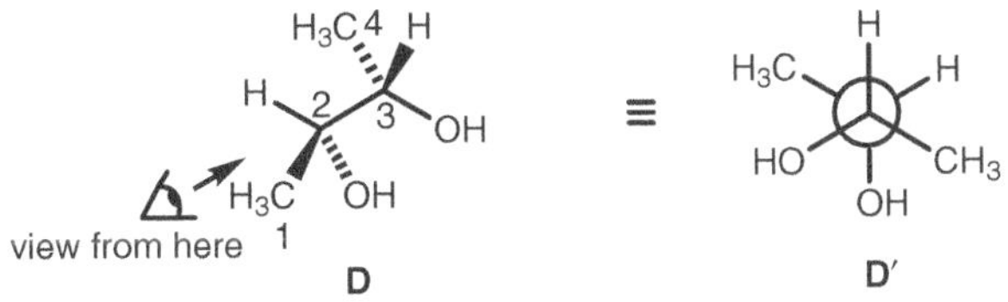

This conformer is the same as **A′** and can be superimposable by an anticlockwise rotation of 120°.

Relative stability C′ > D′ = A′ > B′.

3. (a)

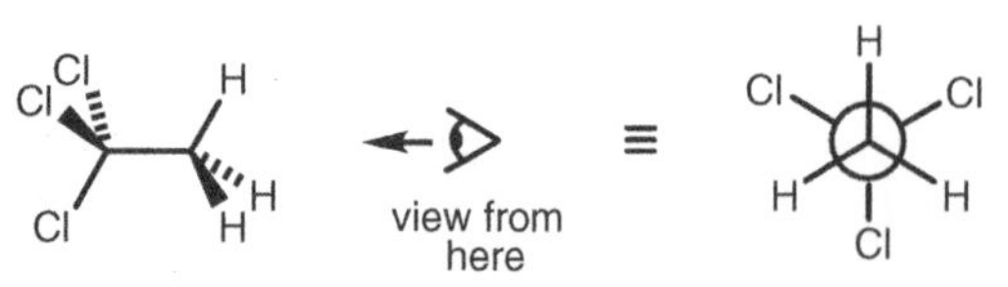

Most stable in a staggered conformation.

(b)

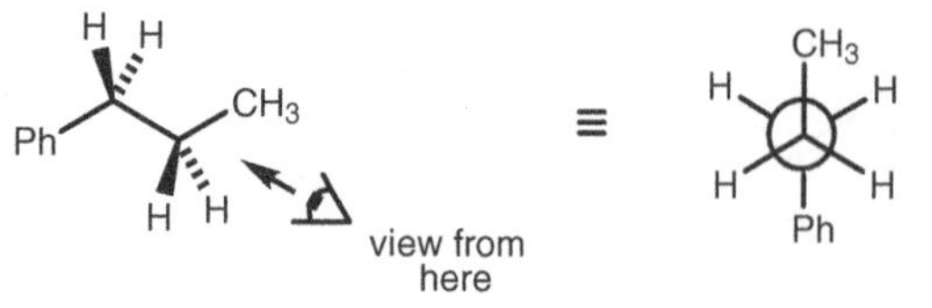

Most stable with antiperiplanar CH_3 and Ph.

(c)

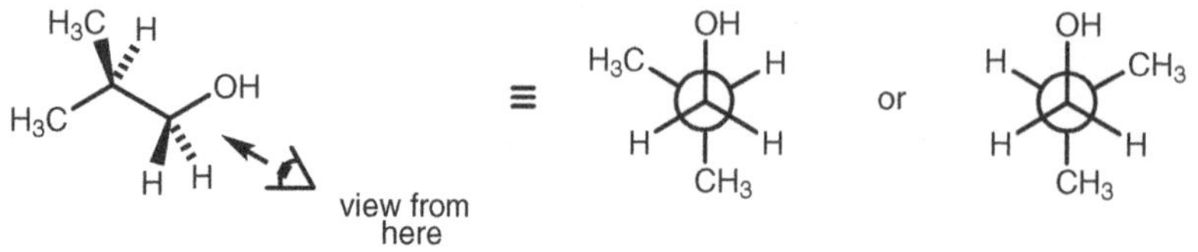

Most stable with small H gauche to both CH_3 groups, larger substituent on the front atom (OH) gauche to only one CH_3. These two enantiomeric conformers have identical energy.

4. These are often easier to see in Newman projections.

(a) $CH_3CH(Cl)CH(Cl)CH_3$: *meso-* has a plane of symmetry.

meso-:

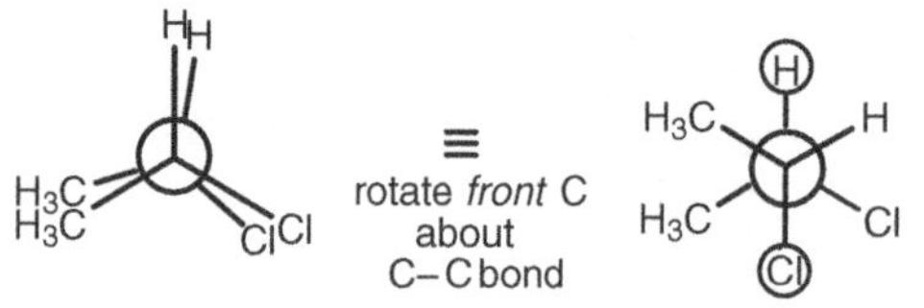

(H) and (Cl) are antiperiplanar.

(b) $CH_3CH(OH)CH(OH)CH_3$: For each C: priority OH>$CHOHCH_3$>CH_3>H

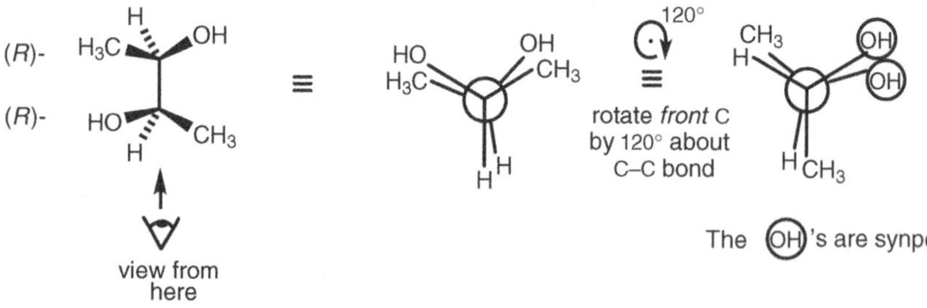

The (OH)'s are synperiplanar.

(c) $CH_3CH(OH)CO_2H$: Priority OH>CO_2H>CH_3>H

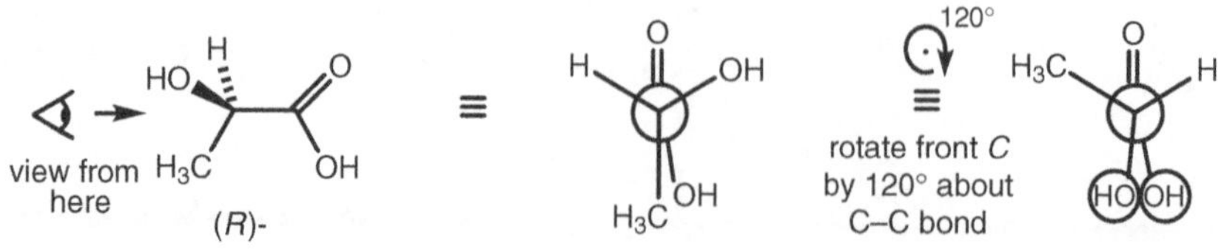

The (OH)'s are synperiplanar.

5.

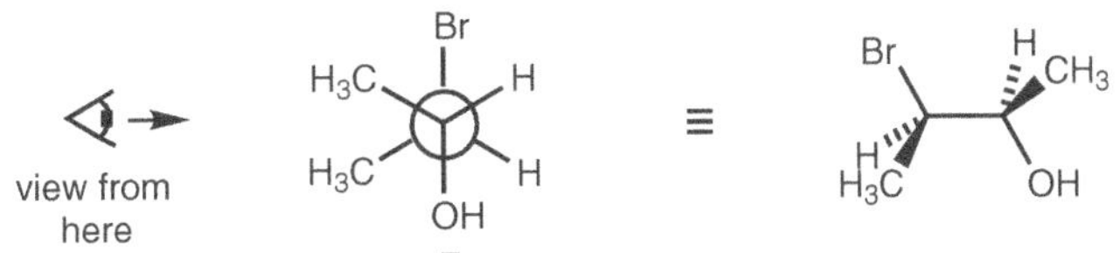

view from here

H₃C Br H, H₃C H, OH

E

≡ Br, H CH₃, H₃C, OH

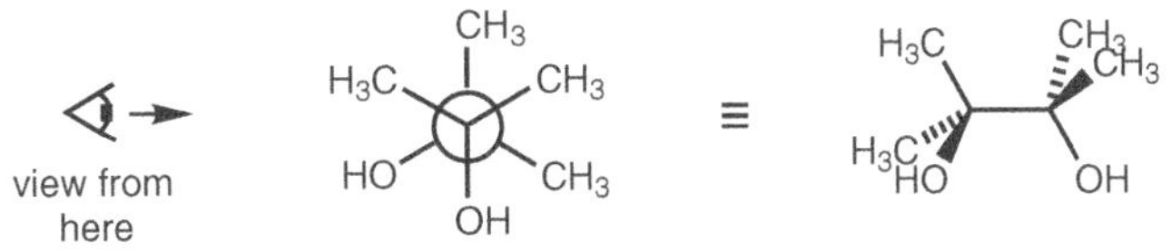

view from here

CH₃, H₃C CH₃, HO CH₃, OH

F

≡ H₃C CH₃ CH₃, H₃C HO OH

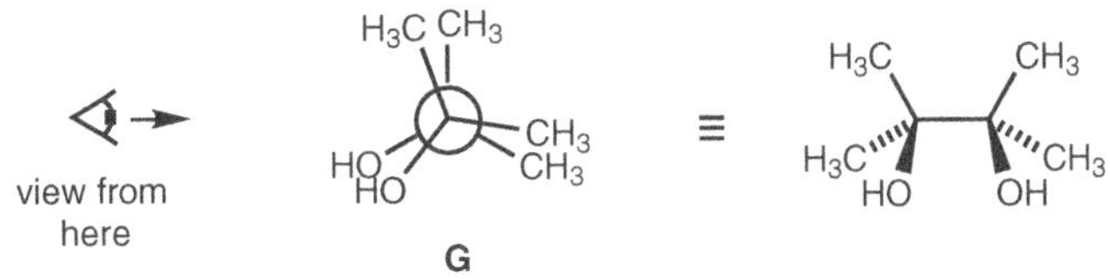

view from here

H₃C CH₃, HO HO CH₃ CH₃

G

≡ H₃C CH₃, H₃C CH₃, HO OH

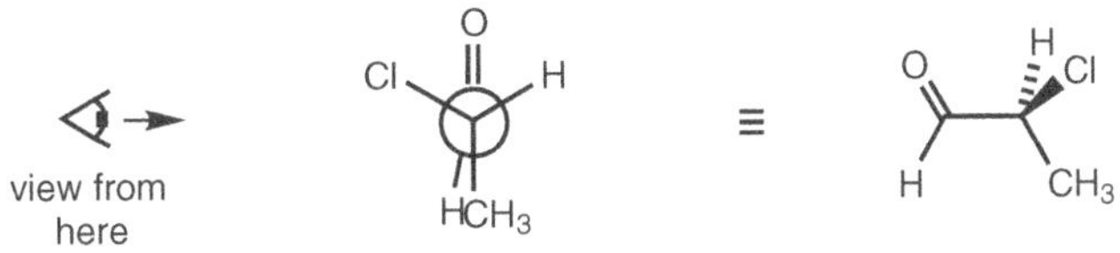

view from here

O, Cl H, HCH₃

H

≡ O, H Cl, H CH₃

6. Draw all the compounds in three dimensions. Put in the important H atoms.

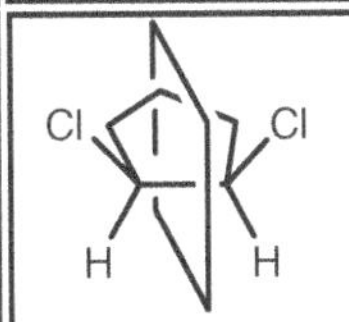

	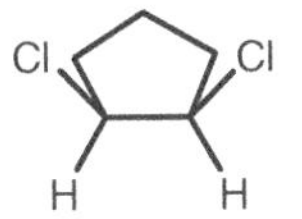 Cl, Cl, H, H (i)	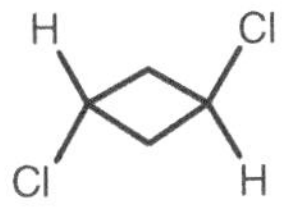 H, Cl, Cl, H (ii)	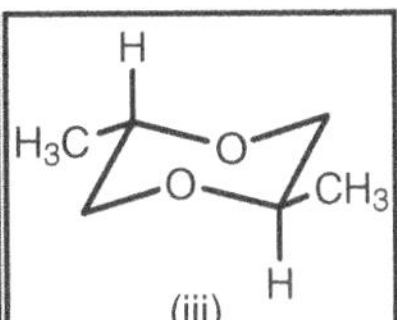 H, H₃C, O, O, CH₃, H (iii)	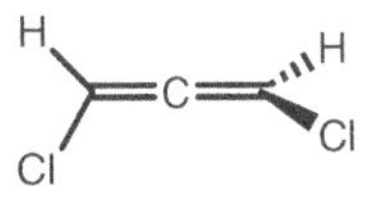 H, H, C, Cl, Cl (iv)
(a) plane? Yes, (i) and (ii)	Cl, Cl, H, H — Yes	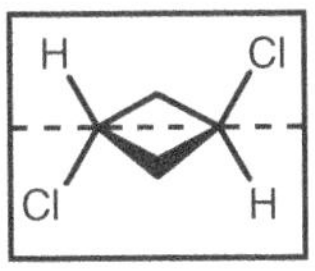 H, Cl, Cl, H — Yes	No	No (The two planar *ends* are at right angles to each other)
(b) centre? Yes, (ii) and (iii)	No	H, Cl, Cl, H — Yes	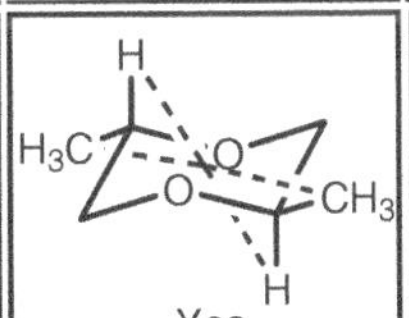 H, H₃C, O, O, CH₃, H — Yes	No
(c) chiral? Only (iv)	No	No	No	Yes

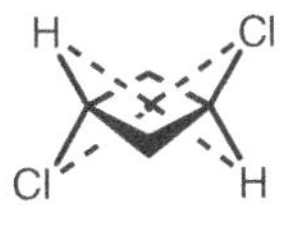

7. Draw the important H atoms. The phenyl group can be written as Ph without losing any stereochemical detail. Assign (R)-/(S)- to the chiral centres.

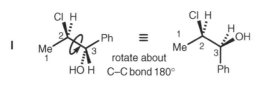

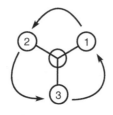

I

at C₂; 2-(R)- - - - - - - - - - - - - - - - - - - at C₃; 3-(S)-

Configuration at C(2)

	Cl	CH(OH)Ph	Me	H
First atom	**Cl**	C	C	**H**
Second atom		**O**,C,H	H,H,H	
Priorities	1	2	3	4

Configuration at C(3)

	OH	CH(Cl)Me	Ph	H
First atom	**O**	C	C	**H**
Second atom		**Cl**,C,H	**C**,C,C	
Priorities	1	2	3	4

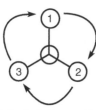

at C₂ clockwise ∴ 2-(R)- at C₃ clockwise ∴ 3-(S)-

J

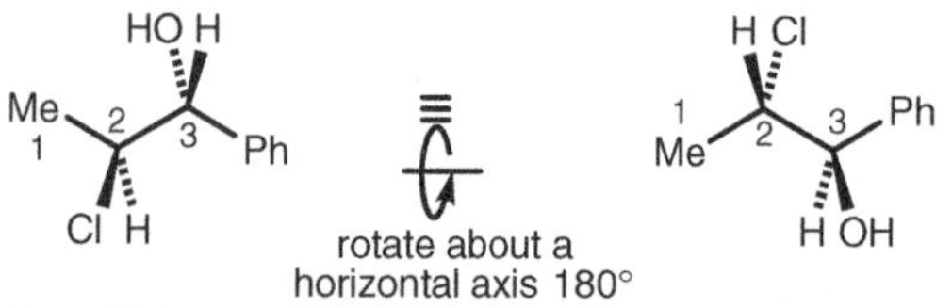

rotate about a
horizontal axis 180°

at C₂; 2-(S)- at C₃; 3-(R)-

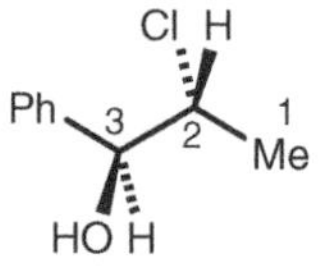

So J is the enantiomer of I

J's non-superimposable
mirror image is (2R,3S) – which is I

K

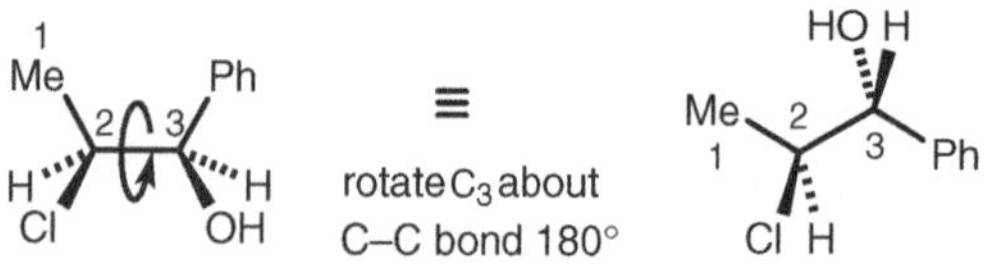

rotate C₃ about
C–C bond 180°

(2S,3R)-

K is the same as J

L

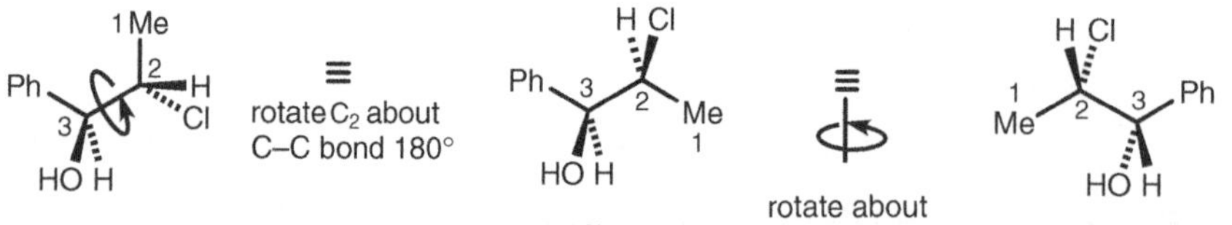

rotate C₂ about
C–C bond 180°

3-(S)-, 2-(S)- rotate about
vertical axis 180° L is a diastereoisomer
of I and of J

8. Draw them all to look as similar as possible, adding important H atoms.
Assign stereochemical descriptors, (R/S)- and (E/Z)-.

(a) They are all stereoisomers of 2-bromo-3-hydroxybutane, which has four stereoisomers, two different enantiomeric pairs: (2S,3S)/(2R,3R) and (2S,3R)/(2R,3S).

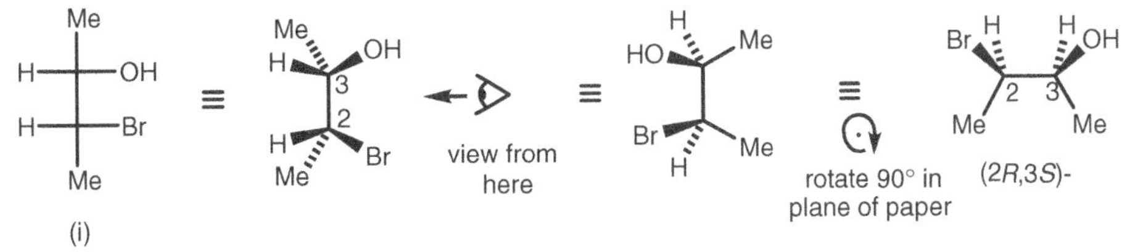

(2R,3S)-

(i)

rotate C₂ about C–C bond 120°   (2S, 3R) – This is the enantiomer of (i).

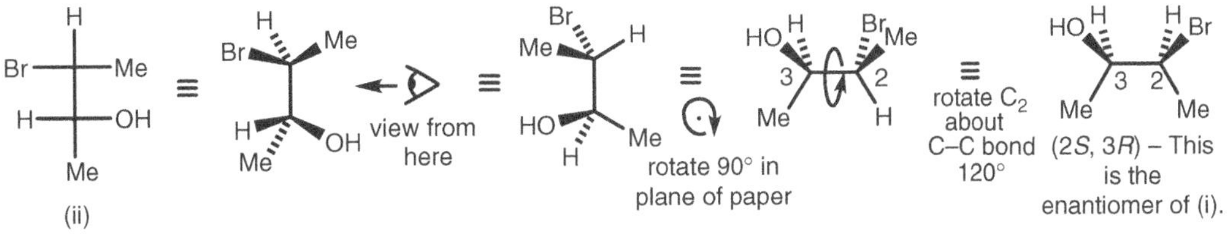

(ii)

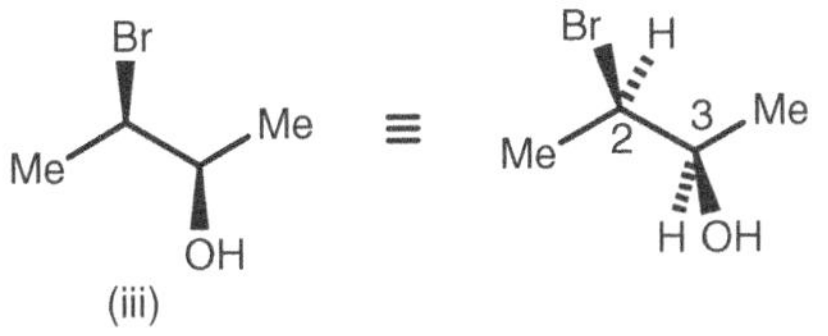

(2R,3R) – a diastereoisomer (epimer) of (i) and of (ii).

(iii)

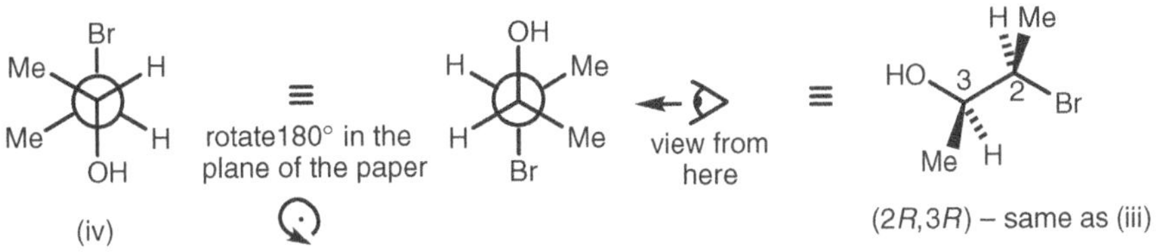

(2R,3R) – same as (iii)

(iv)

(b)   Pr = CH₂CH₂CH₃

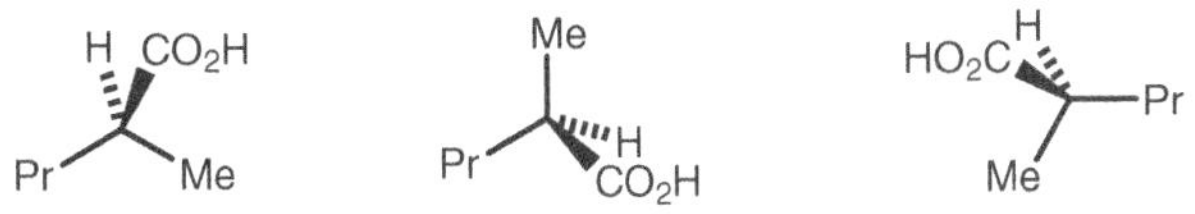

(i) (S)-   (ii) (R)-   (iii) (R)-

All chiral: (ii) and (iii) are identical; (i) is the enantiomer.

(c)

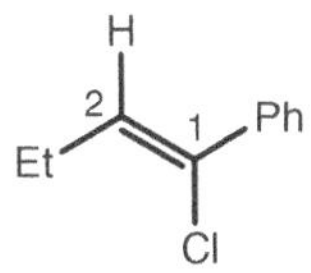

Double bond atom	C(1)	C(2)
Higher priority	**Cl**	**Et**
Lower priority	Ph	H

∴ (Z)-configuration

(i) (Z)-

(ii) (Z)-   (iii) (E)-

All achiral: (i) and (ii) are both (Z); (iii) is the (E)-geometric isomer.

(d)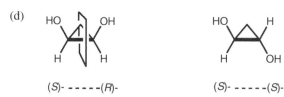

(S)- - - - - -(R)-   (S)- - - - - -(S)-   (R)- - - - - -(R)-

(i) Achiral, *meso-*.   (ii) Chiral (S,S)-diastereoisomer   (iii) Chiral (R,R)-enantiomer of (ii);
Plane of symmetry.   (epimer) of (i).   diastereoisomer (epimer) of (i).

9. Draw all the molecules in three dimensions. Put in the important H atoms.
    (a) two molecules which are achiral and contain no chiral centres;

achiral – superimposable on its mirror image

achiral – superimposable on its mirror image

    (b) two molecules which are achiral and contain chiral centres;

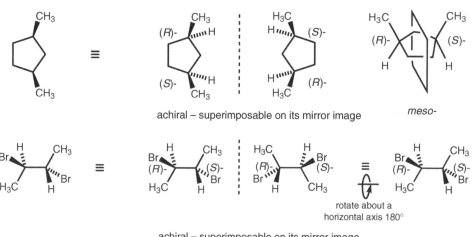

achiral – superimposable on its mirror image   *meso-*

achiral – superimposable on its mirror image

● = centre of symmetry
(or centre of inversion)

meso-   rotate left hand
              carbon 180°

meso- - - - - plane of symmetry

    (c) two molecules which are chiral and contain chiral centres;

chiral – non-superimposable on its mirror image

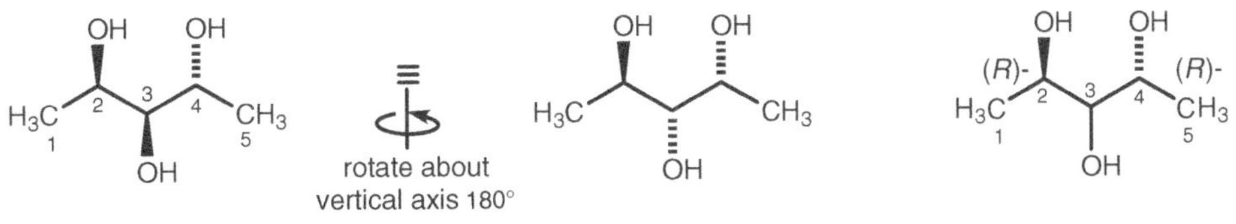

Defining the stereochemistry at the C(3) position causes a problem, since the stereochemistry is neither *up* nor *down*. The stereochemistry at this position is therefore not defined since it is not a chiral centre.

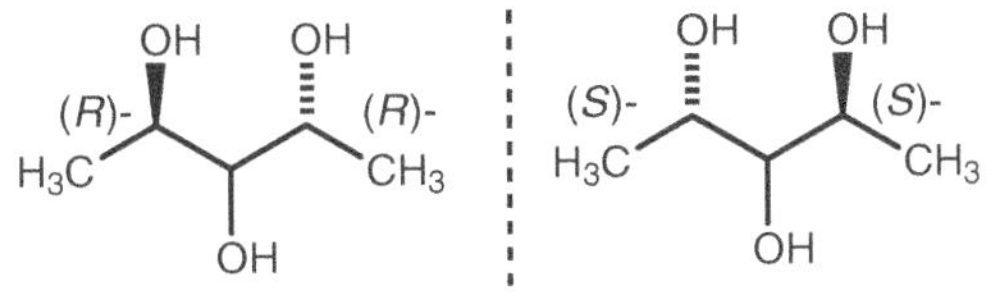

chiral – non-superimposable on its mirror image

(d)   two molecules which are chiral and contain no chiral centres.

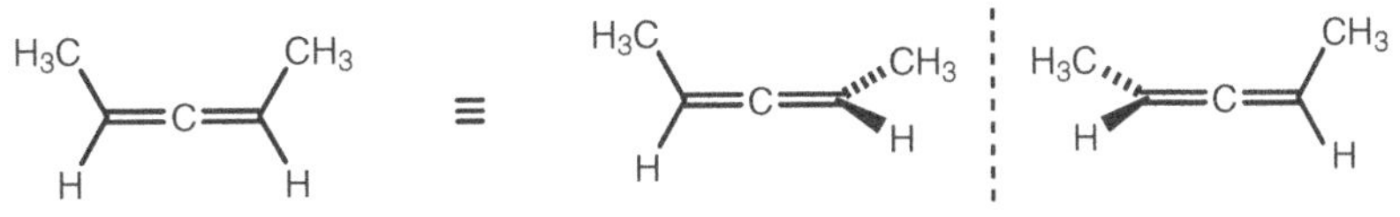

chiral – non-superimposable on its mirror image

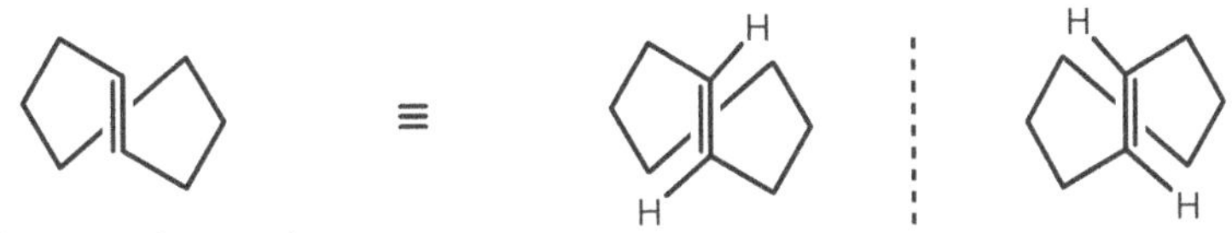

(*trans*-cyclooctene)

chiral – non-superimposable on its mirror image

10.   Look for double bonds, rings and chiral centres. Don't forget to use the correct bond angles and put in any *relevant* hydrogen atoms. Draw any one structure and then look for planes or centres of symmetry; if there are none, draw the mirror image and check to see if it is non-superimposable and thus chiral.

(a)   Has a ring and two chiral centres (marked*), both identically substituted.

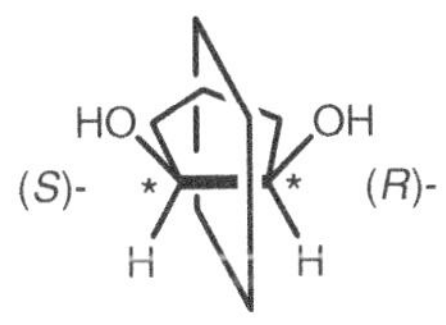

This has a plane of symmetry and is achiral with two chiral centres; it is a *meso*-compound.

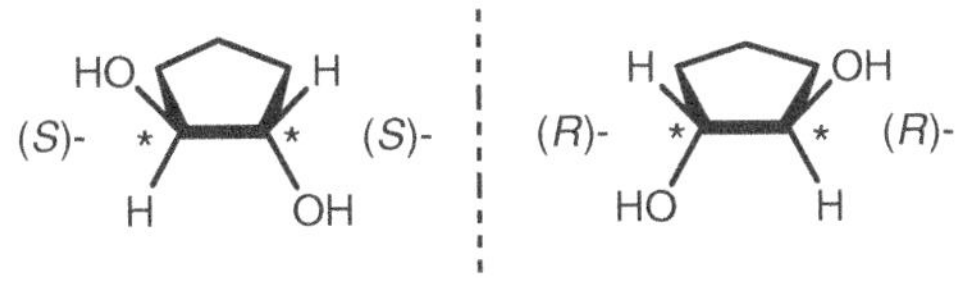

This is a pair of enantiomers (*R,R*)- and (*S,S*)-. They are both diastereoisomeric with the (*S,R*) *meso*-compound, as drawn above.

mirror plane

There are three stereoisomers: one *syn*- (*S,R*) *meso*-compound and an *anti*-enantiomeric pair (*S,S*)- and (*R,R*).[2]

---

[2] The groups that are together are termed *syn*-, and those which are not are termed *anti*- (see abbreviations).

(b) Has one ring and two chiral centres, identically substituted. The amide group is flat.

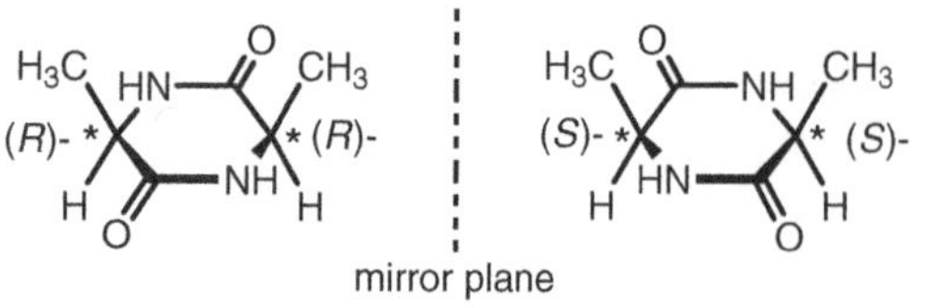

No plane or centre of symmetry; two enantiomers (R,R)- and (S,S).

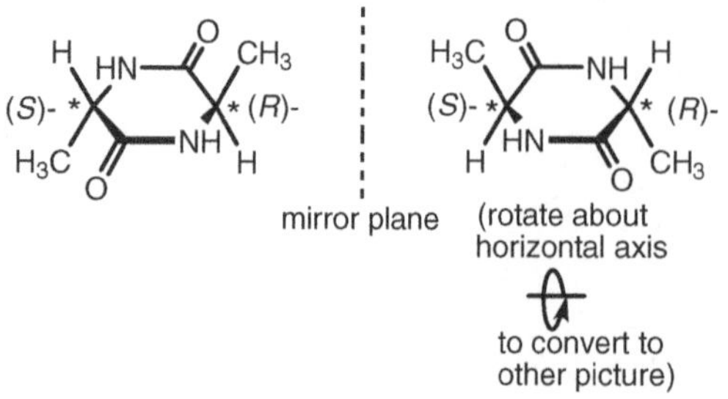

mirror plane

mirror plane

(rotate about horizontal axis

to convert to other picture)

Identical achiral *meso*-structures. (R,S)-diastereoisomeric with the two (R,R)- and (S,S)- enantiomers.

There are three stereoisomers, a *syn*-enantiomeric pair (R,R)- and (S,S)-
and an achiral *anti-meso*-compound (R,S).

(c) Two double bonds; the central C must be sp hybridised to provide two p-orbitals, one each for the two double bonds, which must be at right angles to each other.

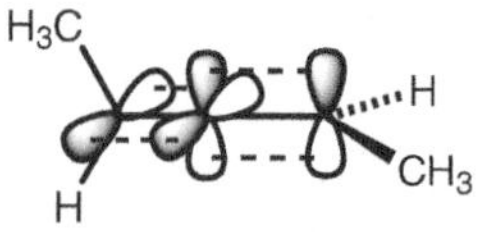

No plane or centre of symmetry, so is there a pair of enantiomers?

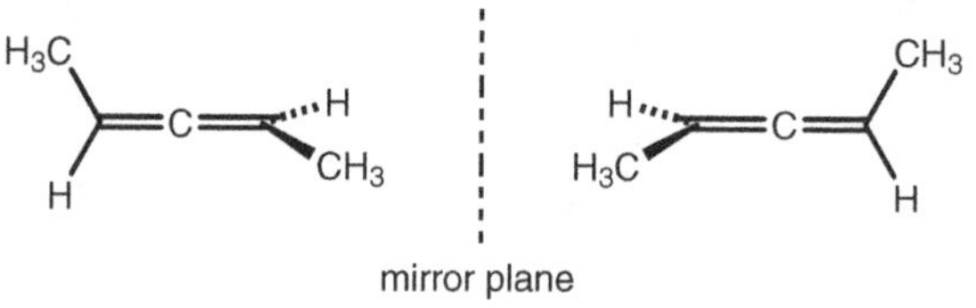

mirror plane

Yes.

(d) One double bond, one ring and one chiral centre.

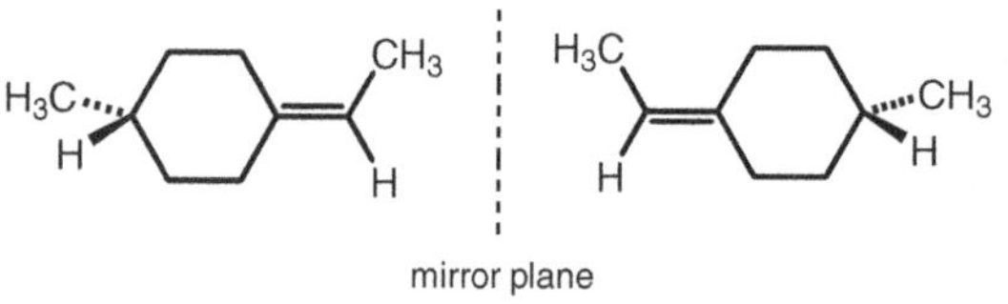

mirror plane

Enantiomers.
To check make molecular models!

(e) Atropisomers

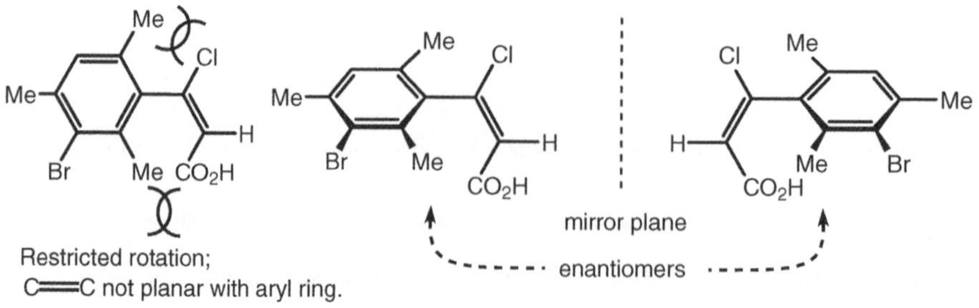

Restricted rotation;
C══C not planar with aryl ring.

mirror plane

enantiomers

(f)   Atropisomers

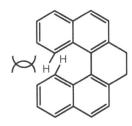 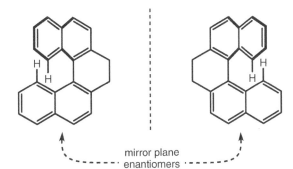

Restricted rotation;
naphthyls not coplanar.
Twisted structure.

mirror plane
enantiomers

(g)   Has one ring, one chiral centre and one double bond, C=N. The N is sp² hydridised and has one non-bonded
      pair of electrons.

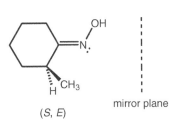

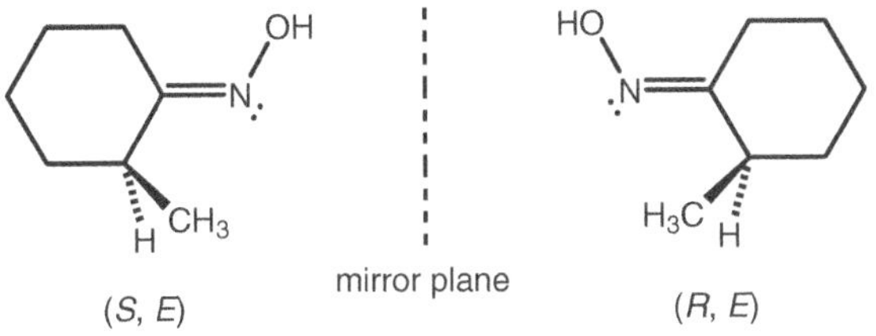

mirror plane

(S, E)   (R, E)

These are enantiomeric (E)-oximes.
They are diastereoisomers of (Z)-oximes.

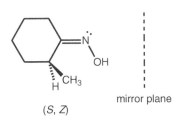

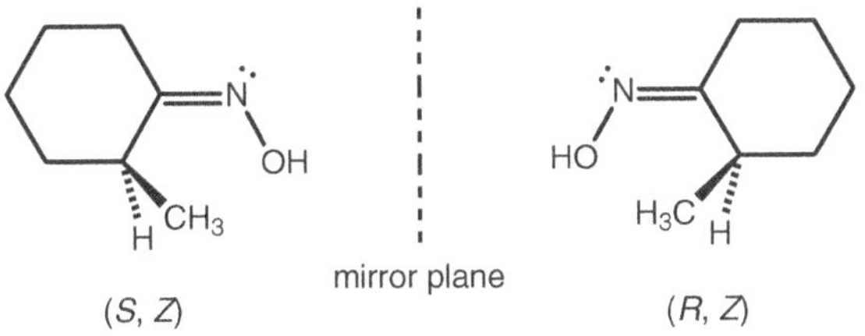

mirror plane

(S, Z)   (R, Z)

These are enantiomeric (Z)-oximes.
They are diastereoisomers of (E)-oximes.

(h)

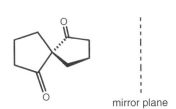

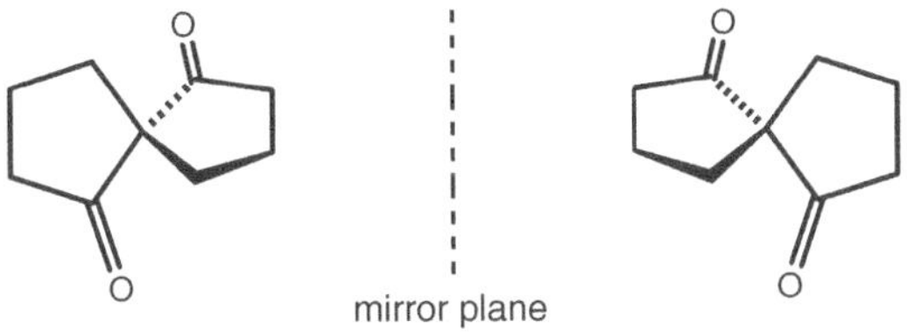

mirror plane

The central C must be tetrahedral, so the
two rings lie at right angles to each other.
These molecules are an enantiomeric pair.

(i)

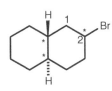

Bromine could be axial or equatorial. No plane or centre of
symmetry. There are three chiral centres marked*.

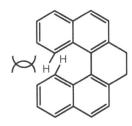

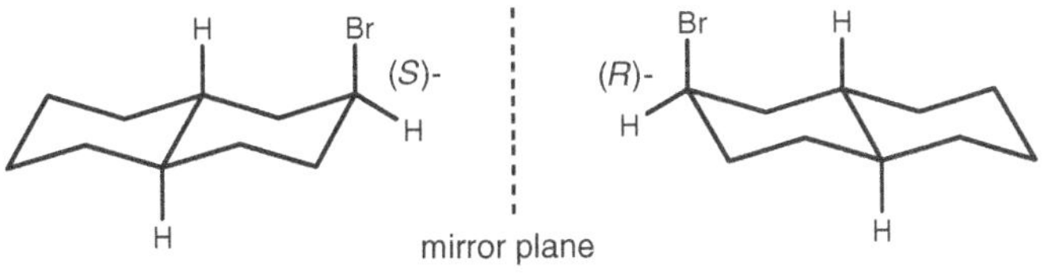

Two enantiomeric axial
bromo compounds.

(S)-   (R)-

mirror plane

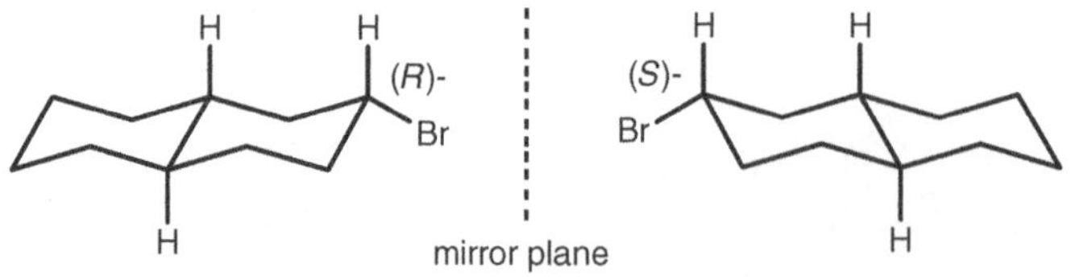

Two enantiomeric equatorial bromo compounds. Both are diastereoisomers of the two axial compounds.

There are four stereoisomers, an enantiomeric axial pair and an enantiomeric equatorial pair.

(j) Restricted rotation; no plane or centre of symmetry in either structure. The aryl rings cannot lie co-planar, so the molecule is twisted.

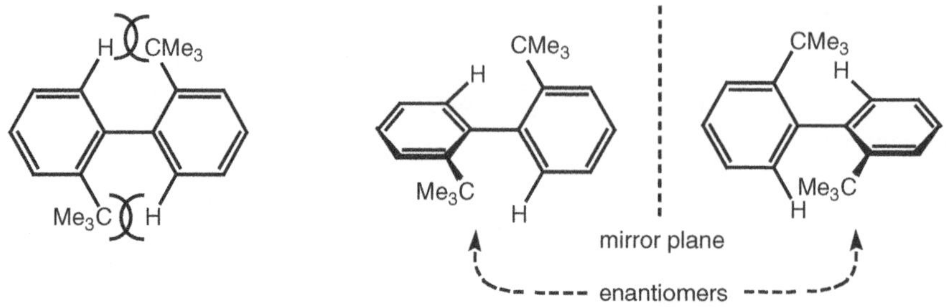

mirror plane

enantiomers

11. For the following molecules:

(i)

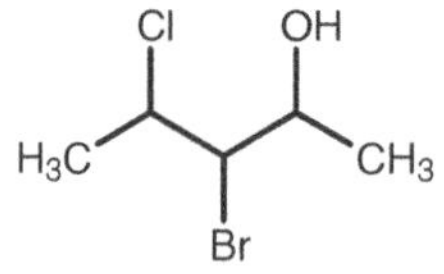

To determine the number of possible stereoisomers:
Firstly, assume all the relative configurations are the same and that the substituents are *coming towards you*; this gives the stereoisomer X_0.

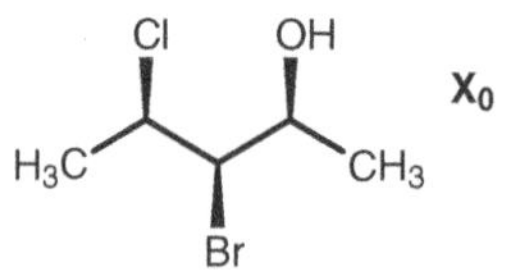

Secondly, change each chiral centre independently of each other by positioning the stereochemistry *away from you* (dashed lines); this gives rise to three additional stereoisomers X_{1a-c}.

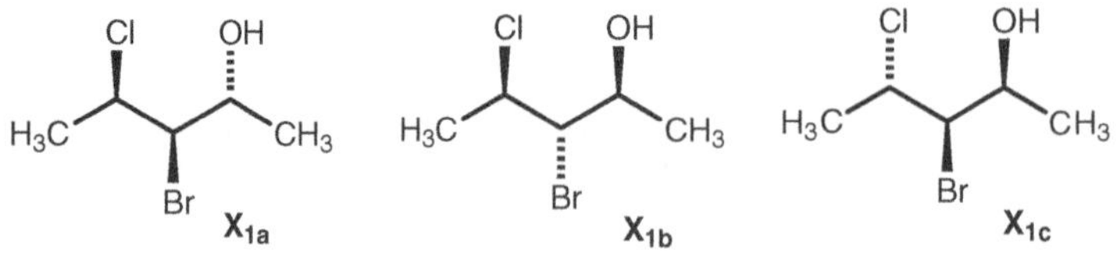

By repeating the procedure again but with any two chiral centres in X_0: this gives rise to three more stereoisomers X_{2a-c}.

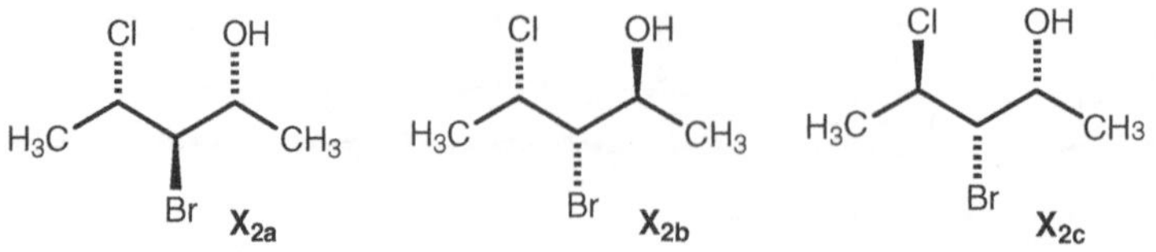

Finally, change all three chiral centres in X_0 to give the final stereoisomer X_3.

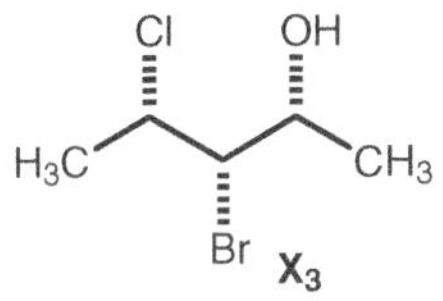

(a) There are eight possible stereoisomers X_0, X_{1a-c}, X_{2a-c} and X_3. This could have also been worked out using the 2^n rule (see page 17). For a molecule with three chiral centres ($n = 3$), the maximum number of possible stereoisomers equals 2^3 ($= 8$ stereoisomers).

(b) There are four pairs of enantiomers (X_0 and X_3, X_{1b} and X_{2a}, X_{1a} and X_{2b}, and X_{1c} and X_{2c}).

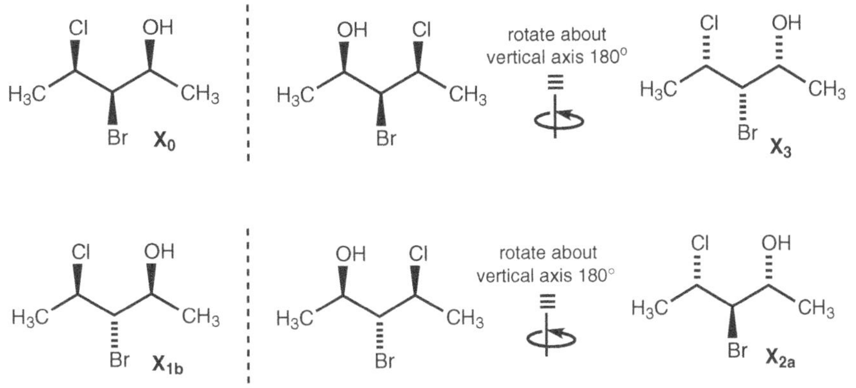

Similarly,

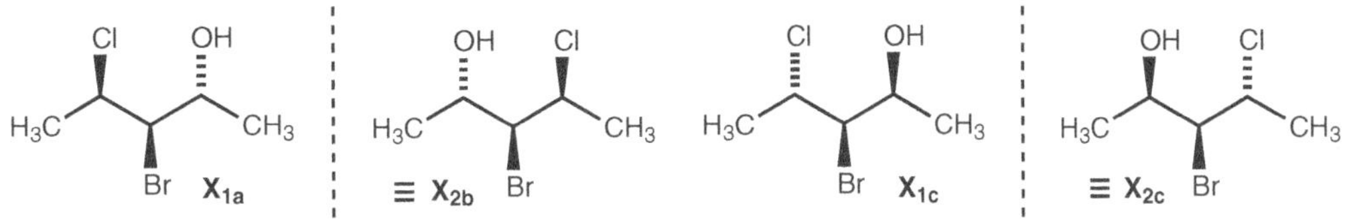

(c) There are six diastereoisomers for each stereoisomer; e.g. for stereoisomer X_0, the six diastereoisomers are X_{1a-c} and X_{2a-c}. The remaining stereoisomer X_3 is its enantiomer.

(d) There are three epimers for each stereoisomer; e.g. for stereoisomer X_0, the three epimer are X_{1a-c}.

(ii)

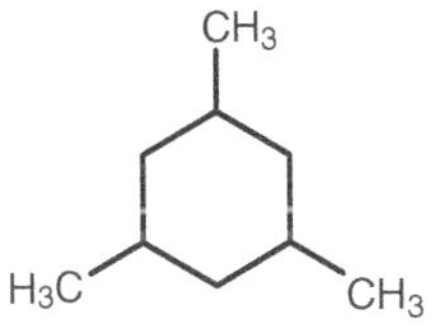

(a) By systematically changing the relative stereochemistry there appear to be eight potential stereoisomers Y_0, Y_{1a-c}, Y_{2a-c} and Y_3.

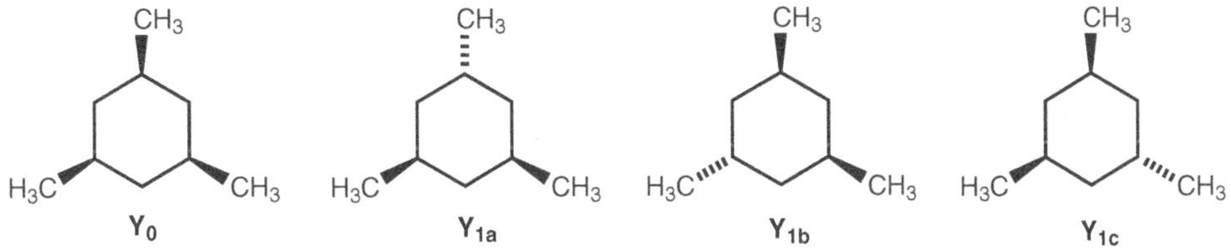

diagrams are continued overleaf

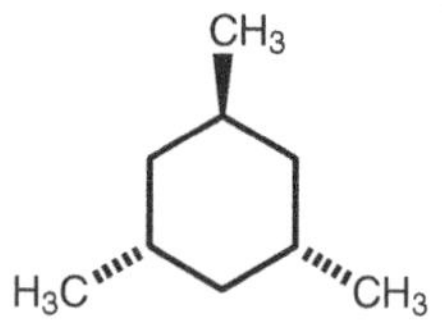

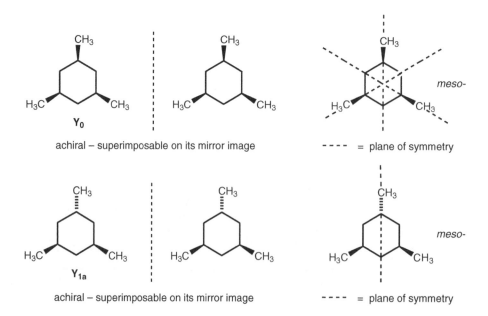

Y$_{2a}$ 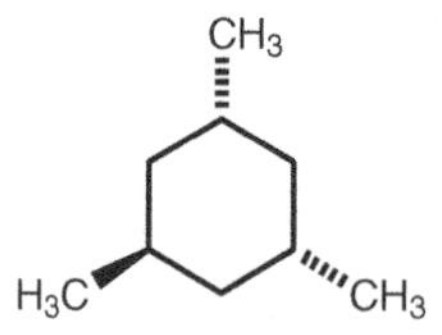   **Y$_{2b}$** 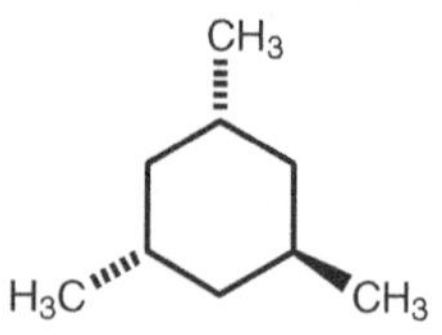   **Y$_{2c}$** 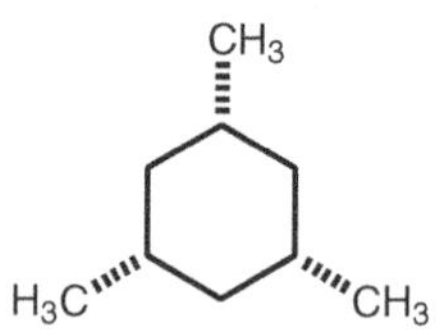   **Y$_3$**

However,

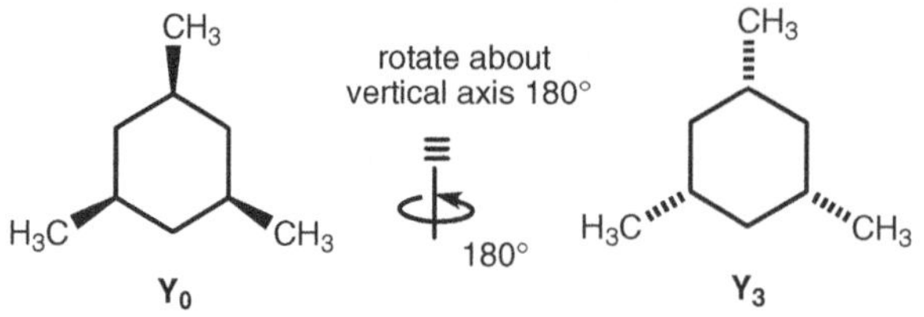

Y$_0$ rotate about vertical axis 180°   180°   **Y$_3$**

and

Y$_{1a}$ rotate 120° in plane of paper   120°   **Y$_{1b}$**   rotate 120° in plane of paper   120°   **Y$_{1c}$**

III   rotate about vertical axis 180°

Y$_{2a}$ rotate 120° in plane of paper   120°   **Y$_{2b}$**   rotate 120° in plane of paper   120°   **Y$_{2c}$**

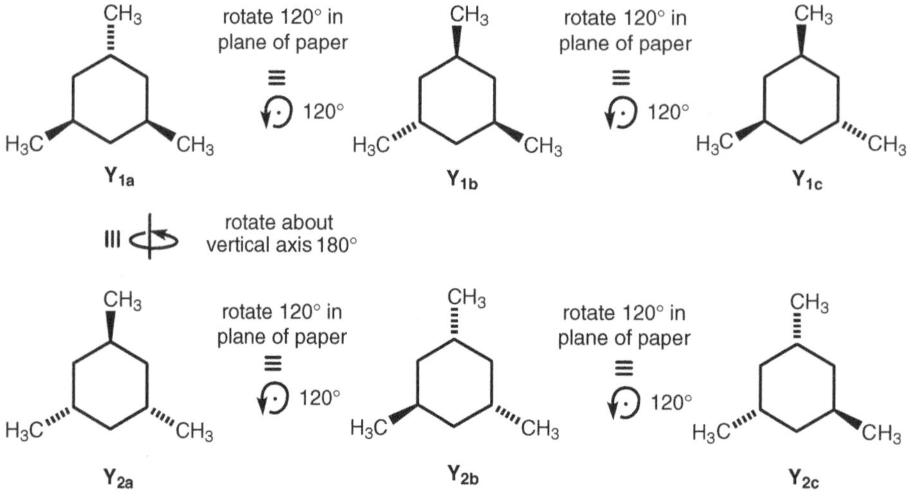

From the eight pictorial representations **Y$_0$**, **Y$_{1a-c}$**, **Y$_{2a-c}$** and **Y$_3$**, there are only two stereoisomers **Y$_0$** (=**Y$_3$**) and **Y$_{1a}$** (=**Y$_{1b}$**, **Y$_{1c}$**, **Y$_{2a}$**, **Y$_{2b}$** and **Y$_{2c}$**). The overall number of stereoisomers has been reduced from the theoretical maximum number of eight to two due to the symmetry present within these molecules.

(b)   There are no pairs of enantiomers. Both stereoisomers **Y$_0$** and **Y$_{1a}$** are *meso*-.

Y$_0$ *meso*-

achiral – superimposable on its mirror image   - - - - = plane of symmetry

Y$_{1a}$ *meso*-

achiral – superimposable on its mirror image   - - - - = plane of symmetry

(c) There are two diastereoisomers. Both diastereoisomers Y_0 and Y_{1a} are *meso-*.

(d) There is a pair of epimers. Both epimers Y_0 and Y_{1a} are *meso-*.

(iii)

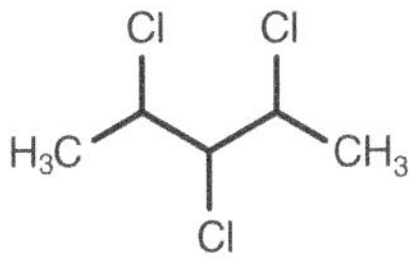

By systematically changing the relative stereochemistry there appears to be eight potential stereoisomers (Z_0, Z_{1a-c}, Z_{2a-c} and Z_3).

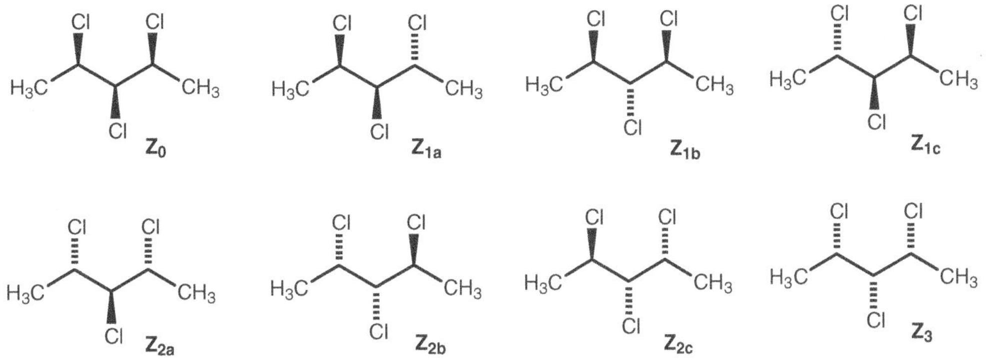

(a) There are eight pictorial stereoisomers Z_0, Z_{1a-c}, Z_{2a-c} and Z_3. However, the overall number is reduced to four due to the symmetry present within these molecules; namely Z_0 ($=Z_3$), Z_{1a} ($=Z_{2c}$), Z_{1b} ($=Z_{2a}$) and Z_{1c} ($=Z_{2b}$).

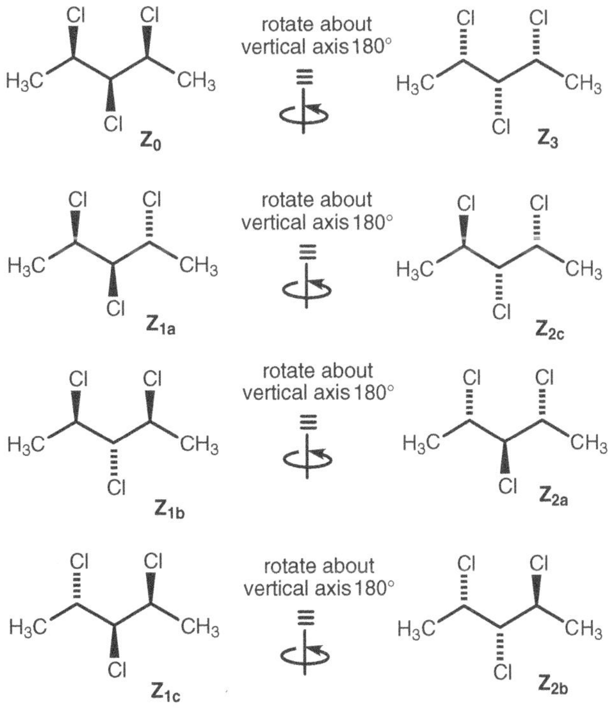

(b) There is a pair of enantiomers; namely, Z_{1a} and Z_{1c}. The remaining stereoisomers Z_0 and Z_{1b} are *meso-*.

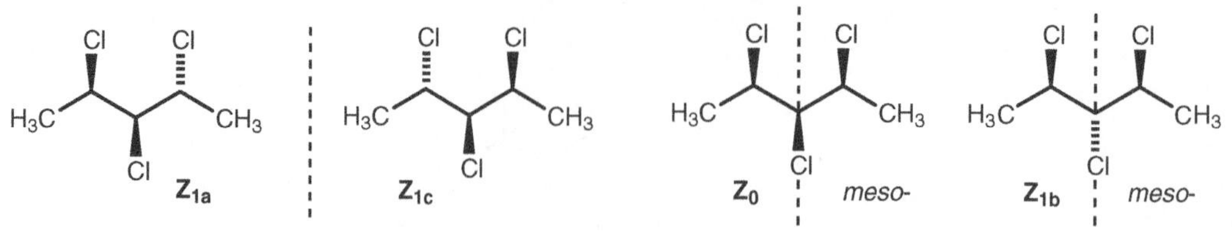

chiral – non-superimposable mirror image

- - - - = plane of symmetry

(c) For stereoisomer Z_0, there are three diastereoisomers (Z_{1a}, Z_{1b} and Z_{1c}).

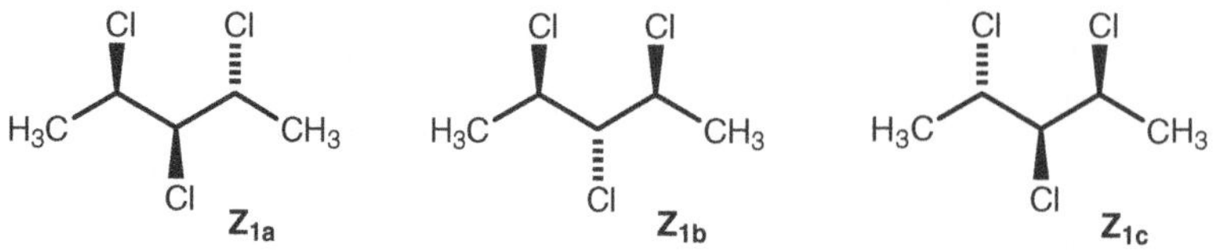

For stereoisomer Z_{1a}, there are two diastereoisomers (Z_0 and Z_{1b}).

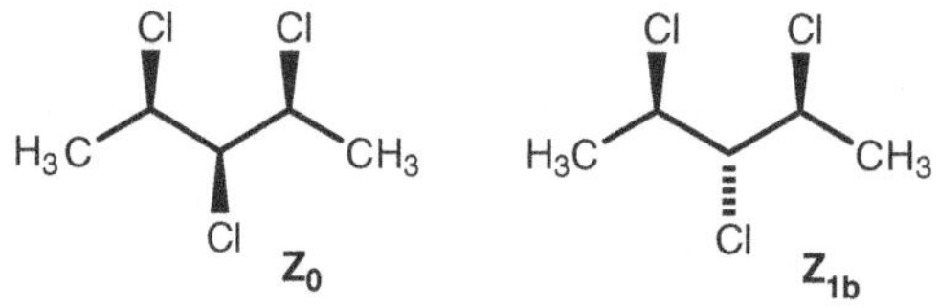

For stereoisomer Z_{1b}, there are three diastereoisomers (Z_0, Z_{1a} and Z_{1c}).

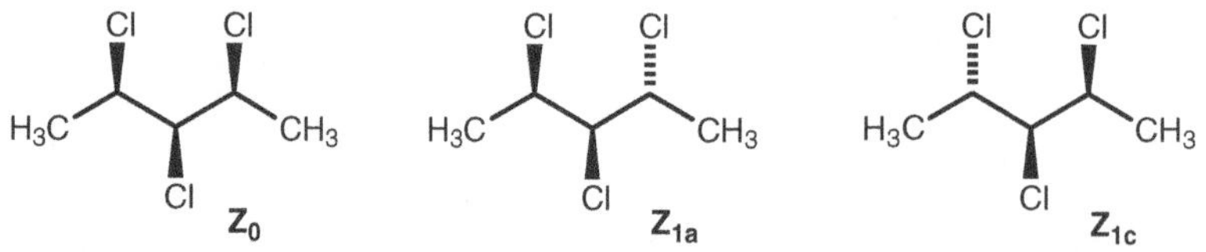

For stereoisomer Z_{1c}, there are two diastereoisomers (Z_0 and Z_{1b}).

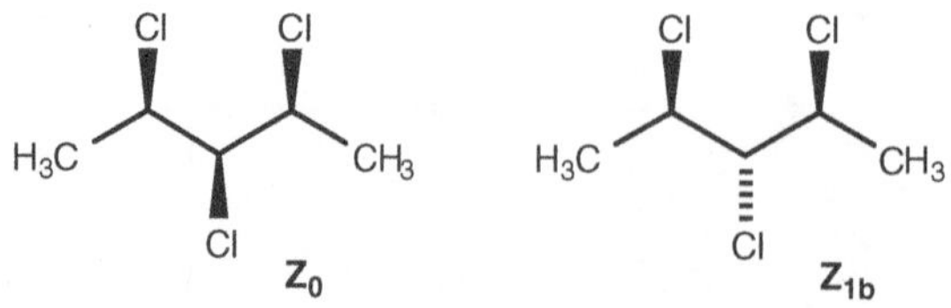

(d) Each stereoisomer has three epimers.

Additional problems for Chapter 1

1. Assign the configuration(s) for the following molecules.

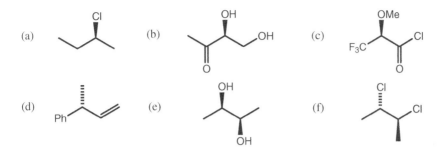

2. Which of the following molecules have the same configuration and conformation?

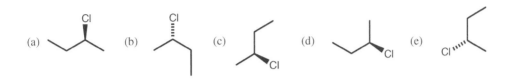

3. Predict the number of possible stereoisomers for the following molecules:
   (a) MeCH(OH)CH(Cl)Et
   (b) MeCH=CH−CH(Me)CO₂H
   (c) HC≡C−CH=CH−CH=CH−CH(OH)Me
   (d) PhCH=C=CH−CH₂−CH=CH−CH(NH₂)Me
   (e) Ph(SO)CH₂CH=CH−CO₂H
4. Draw in three dimensions all the possible stereoisomers for the following molecules. Predict if they can be obtained optically pure at room temperature.

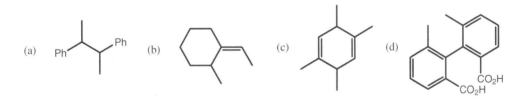

5. From the following molecules, pick out:
   (a) two with a plane of symmetry;
   (b) two with a centre of symmetry;
   (c) one that is chiral.

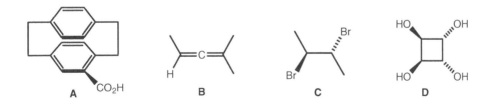

Answers: 1. (a) (*S*)-; (b) (*S*)-; (c) (*R*)-; (d) (*R*)-; (e) (*R,R*)- and (f) (*S,S*)-; 2. (a), (b), (c) and (e) have the same (*S*)-configuration and (d) has an (*R*)-configuration. (a) + (c), and (b) + (e) have the same conformation and configuration. (b) and (d) have the same conformation, but a different configuration; 3. (a) (*R/S*) × (*R/S*) = 4 stereoisomers; (b) (*E/Z*) × (*R/S*) = 4 stereoisomers; (c) (*E/Z*) × (*E/Z*) × (*R/S*) = 8 stereoisomers; (d) (*R/S*) × (*E/Z*) × (*R/S*) = 8 stereoisomers; (e) contains a chiral sulfur atom; (*R/S*) × (*E/Z*) = 4 stereoisomers; 4. (a) there are 3 stereoisomers (2 enantiomers + 1 *meso*); (b) there are 2 stereoisomers (2 enantiomers); (c) there are 3 stereoisomers (2 enantiomers + 1 *meso*); (d) there are 2 stereoisomers (2 atropisomeric enantiomers – see pages 14, 32 and 33); 5. (a) **B** and **D**; (b) **C** and **D**; (c) **A**.

2 Nucleophilic Substitution Reactions of Saturated Carbon Species

The aim of this chapter is to develop an understanding of the stereochemical features of nucleophilic substitution reactions at sp^3 hybridised carbon atoms.

At the end of this chapter, you should be able to:

- Draw and compare the mechanisms for S_N2, S_N2', S_N1 and S_Ni reactions, and recognise the differences between them.
- Appreciate the competition between substitution and elimination reactions, and the differences between nucleophiles and bases.
- Predict, explain and use the stereochemical consequences of these reactions.
- Explain the influence of neighbouring groups on these reactions, and comment on the stereochemical outcome.

Background: substitution reactions

Formally these reactions may be divided into two major groups.

S_N2: Single stage, bimolecular reaction with no intermediates;
S_N1: Two-stage, unimolecular reaction with carbocation intermediates.

These two substitution mechanisms actually describe the ends of a continuous spectrum of mechanisms which vary according to the extent to which bond breakage to the leaving group occurs before bond making by the incoming nucleophile.

The *nucleophiles* involved are species with high-energy occupied molecular orbitals (HOMOs) such as a cyanide ion, ammonia or arenes. Non-bonded electron pairs or electron-rich π-orbitals make good HOMOs.

The *electrophiles* with which they react can be sp^3 hybridised carbon atoms bearing polar bonds to good leaving groups, such as $^-OSO_2Ar$, Br^-, and $-O^-$ in epoxides. The polar bonds have low-energy unoccupied σ^* molecular orbitals (LUMOs). The vacant orbital of a carbocation also makes an excellent LUMO.

S_N2 mechanism

The in-line stereochemistry of the transition state for an S_N2 reaction is dictated by the 180° angle between the σ- and σ^*- orbitals of the sp^3 carbon atom. This results in inversion of configuration at a chiral (stereogenic) centre. In the transition state, the carbon and its other three substituents take up a planar arrangement (with bond angles of approximately 120°) at right angles to the Nu-C-LG axis.

HOMO
non-bonded
pair

LUMO
σ^* of C–LG bond

Trigonal bipyramidal
transition state

Inversion of configuration is observed when
the starting material is chiral.

(where Nu = nucleophile and LG = leaving group)

Neighbouring group effects

A single S_N2 reaction causes inversion, but two will lead to overall retention of configuration. Such a double inversion can occur by participation of a *neighbouring group* within the molecule, which initially acts as a nucleophile in the first stage and then as the leaving group in the second. The molecularity of the reaction will depend upon which stage is rate limiting.

(where NG = neighbouring group)

Because nucleophiles are also bases, E2 eliminations can compete with S_N2 substititions (see Chapter 3 for eliminations).

S_N2' mechanism

This is a bimolecular substitution reaction on an allylic compound which occurs with simultaneous rearrangement of the double bond.

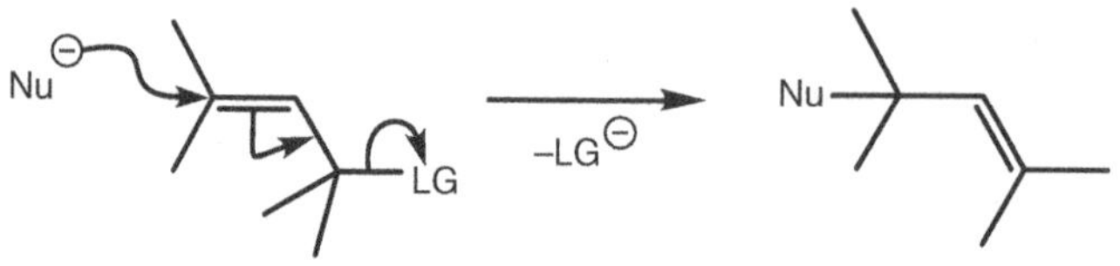

The leaving group departs from the same face of the molecule as the one the nucleophile enters. The electrons flow from the nucleophile through the double bond to eject the leaving group at the end, using the overlap of the π-system and the σ* of the C–LG bond.

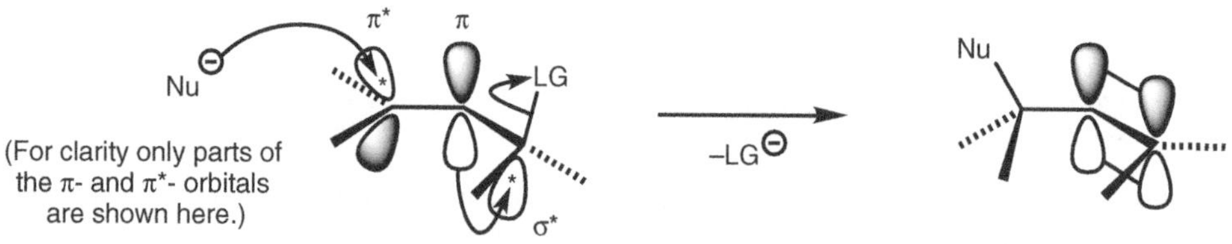

(For clarity only parts of the π- and π*- orbitals are shown here.)

Allylic compounds can also react by an S_N2 mechanism (without rearrangement) and by an S_N1 mechanism.

S_N1 mechanism

The intermediate here is a planar, trigonal sp^2 carbocation having the vacant p-orbital at right angles to the plane. Attack on either face of the symmetrical, planar cation should be equally easy, leading to a racemic product from an enantiomerically pure starting material.

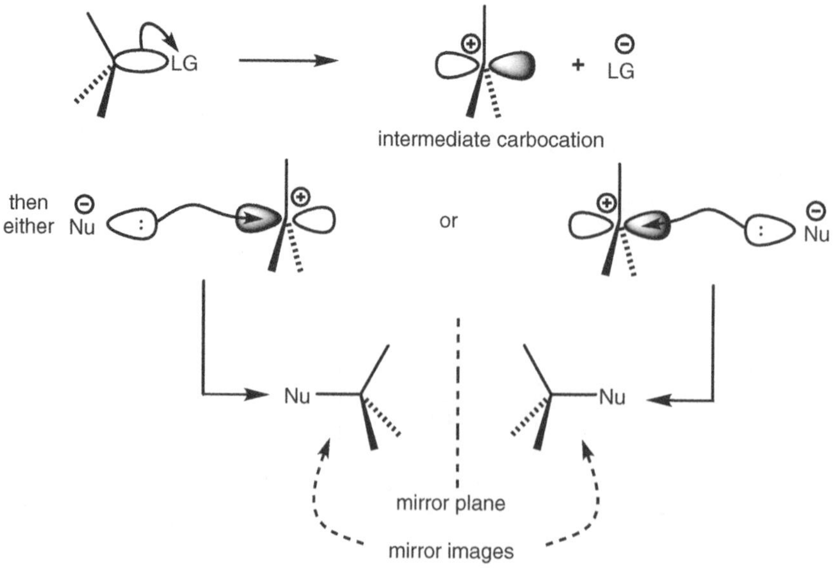

intermediate carbocation

then either

or

mirror plane

mirror images

The carbocation is also an intermediate in E1 eliminations, which can therefore compete with S_N1 substitions (see Chapter 3 for E1 eliminations).

S_N_i mechanism

There is also a small group of *internal* substitutions which characteristically give retention of configuration. Some of these reactions have been shown to proceed via a short-lived close ion pair, where part of the leaving group (e.g. Cl⁻) can react with the intermediate cation on the *same face* from which it was released (e.g. by decomposition of a chlorosulfite ester). If the cation lives long enough for the ions to diffuse away, then the S_N1 mechanism (racemisation) takes over. If the solvent is a good nucleophile, the mechanism reverts to S_N2.

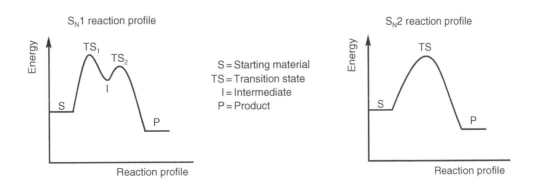

This unimolecular conversion of a chlorosulfite ester, ROSOCl, into RCl and SO_2 can also occur via solvent-participation (e.g. diethyl ether); this would involve two consecutive S_N2 reactions and as a result can lead to retention of configuration (see neighbouring group participation on page 42).

Choosing between S_N_1 and S_N_2

One way to approach the problem of assigning a mechanism to a given substitution reaction is to assume that it will be S_N2 unless there are good reasons to think otherwise. For S_N1, look for a good leaving group and a stabilised carbocation intermediate, preferably delocalised, with a polar solvent which can stabilise ions (especially if the solvent can form hydrogen bonds with the departing leaving group). If a favourable carbocation is not available, the reaction will more than likely proceed via an S_N2 mechanism, especially if there is a good nucleophile present; a dipolar aprotic solvent like dimethyl formamide (DMF) or dimethyl sulfoxide (DMSO) helps too. It is also important to note the difference between an intermediate and a transition state within these reactions; an intermediate I occurs at an energy minimum at the bottom of the ∪ shape present in a reaction energy profile diagram, whereas, a transition state TS occurs at an energy maximum at the top of the ∩ shape.

S_N_1 reaction profile

S_N_2 reaction profile

S = Starting material
TS = Transition state
I = Intermediate
P = Product

1. Suggest reagents and propose a mechanism for the following conversion.

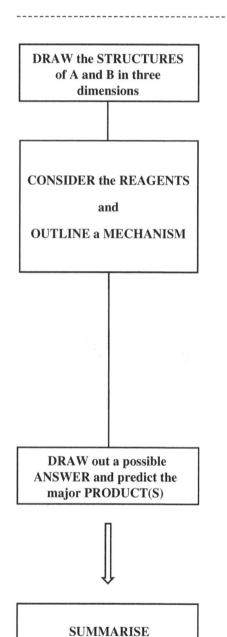

One chiral centre is present in both **A** and **B**.
Inspection of the 3-D drawings shows that inversion of stereochemistry is needed.

OH is lost and CN is gained. For ROH to RCN: ⁻OH is a poor leaving group, so an intermediate with a better leaving group is needed. To achieve an overall inversion in two reactions, you need *one* inversion and *one* retention.

TsO⁻ is a good leaving group. TsCl in the presence of pyridine is used for the conversion of ROH into ROTs; this does not affect the chiral centre. Cyanide ion would then displace ⁻OTs by an S_N2 mechanism (with inversion). Alternatively, conversion of –OH into –Cl by an S_Ni reaction (SOCl₂, with retention see page 43) followed by KCN (S_N2, inversion) would also work.

Note: ROH + KCN   *will not work:*

ROH + ⁻CN ⇌ RO⁻ + HCN

Nor will ROH + HCN: HCN is a weak acid, ROH is a weak base so there will be almost no protonation of ROH by HCN. There is also very little free ⁻CN, hence no reaction!

When you have finished drawing the mechanism, re-draw your product to check that the stereochemistry is correct.
When drawing the stereochemistry for an S_N2 reaction, it often helps to draw the starting molecule so that the bond from C to the leaving group is in the plane of the paper. The bond formed to the incoming nucleophile will then also be in the plane of the paper (which is much easier to see) while the other three groups on the carbon turn inside-out like an umbrella in the wind.

Reagents
(i)   TsCl/pyridine (base + nucleophilic catalyst).
(ii) KCN.

The intermediate tosylate (sulfonate ester) is formed with retention of configuration by reaction with TsCl in pyridine. This is converted to the required (*S*)-nitrile **B** via an S_N2 displacement by ⁻CN (from KCN) with inversion of configuration.

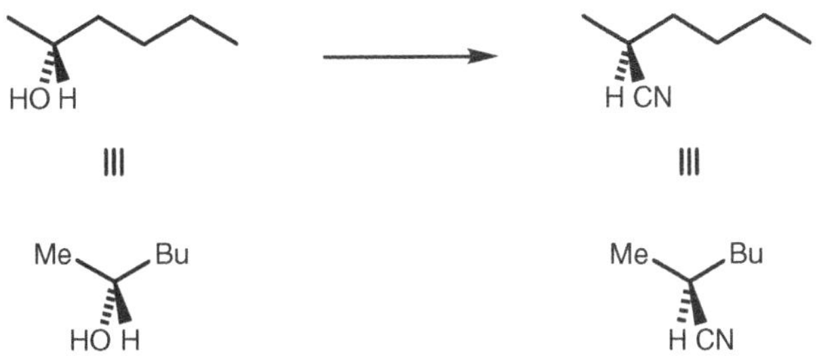
HO H
H CN
Me
Bu
HO H
Me
Bu
H CN

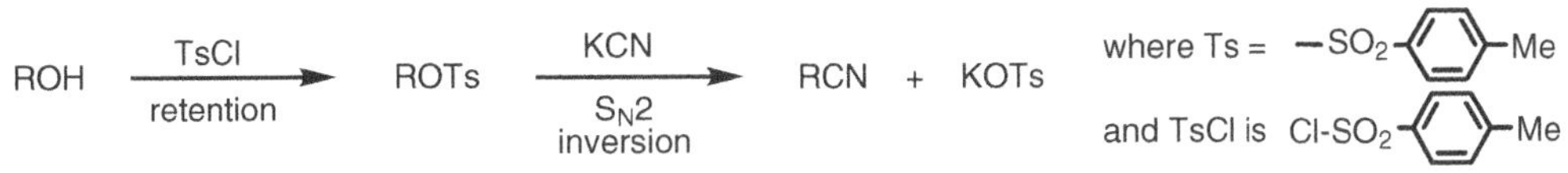
ROH
TsCl
retention
ROTs
KCN
SN2
inversion
RCN + KOTs
where Ts = —SO2-⟨⟩-Me
and TsCl is Cl-SO2-⟨⟩-Me

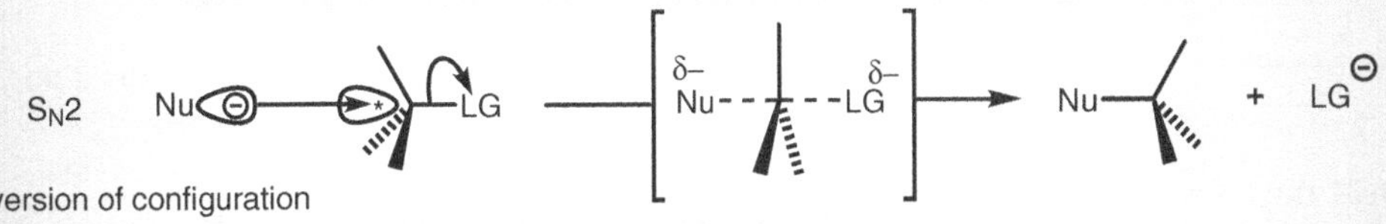
SN2
Nu
LG
δ–
Nu
δ–
LG
Nu
+ LG

Inversion of configuration

HOMO: non-bonded pair of electrons, Nu^-; LUMO: σ* of C–LG

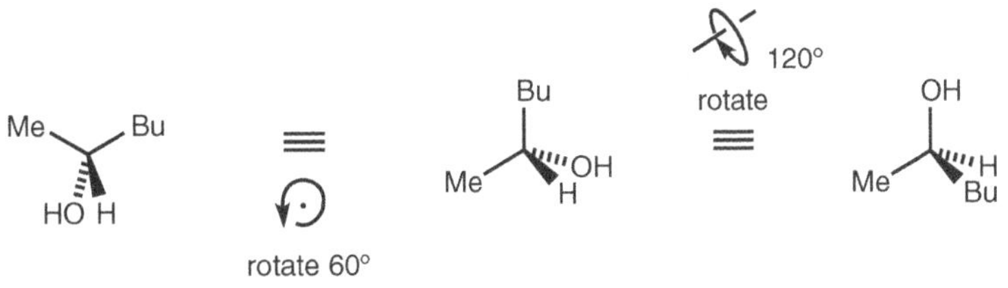
Me
Bu
HO H
rotate 60°
Bu
Me
OH
H
120°
rotate
OH
Me
H
Bu

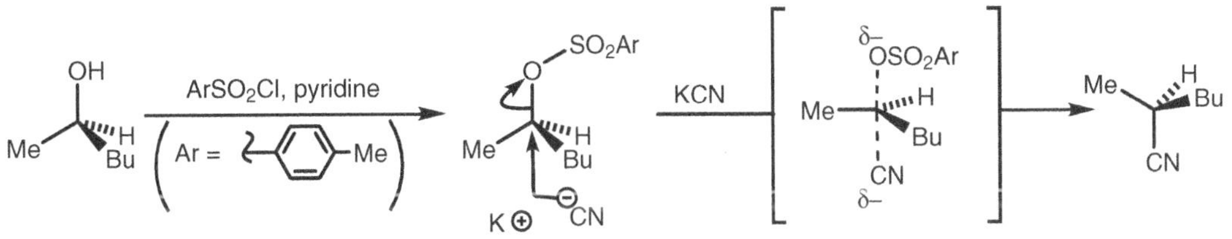
OH
Me
H
Bu
ArSO2Cl, pyridine
Ar =
Me
SO2Ar
O
Me
H
Bu
K
CN
KCN
δ–
OSO2Ar
Me
H
Bu
CN
δ–
Me
H
Bu
CN

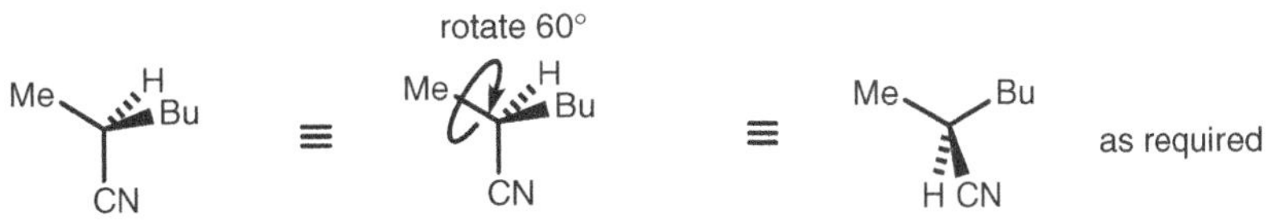
Me
H
Bu
CN
rotate 60°
Me
H
Bu
CN
Me
Bu
H CN
as required

2. Optically active (*S*)-2-iodohexane undergoes racemisation when treated with NaI in propanone. The rate of this reaction depends on the concentration of both iodohexane and NaI. Explain these observations.

DRAW the STRUCTURE of (S)-iodohexane

(*S*)-2-Iodohexane has one chiral centre. The starting material is a single enantiomer, but the product is racemic. Therefore the reaction must pass through a symmetrical intermediate or transition state.

CONSIDER the REAGENT

and

OUTLINE a MECHANISM

Iodine ion can act as both the nucleophile and leaving group.
It is a good nucleophile because it is large and easily polarised, but it is also a good leaving group because it is weakly bonded and forms a stable anion due to minimal lone pair–lone pair repulsion.

The kinetics suggests at least a bimolecular process, such as an S_N2 reaction, with rate $\propto$ [RI][I$^-$].

The rate of a unimolecular S_N1 substitution would be proportional to [RI] only.

DRAW out a possible ANSWER and predict the major PRODUCT(S)

An S_N2 reaction on each (*R*)-iodohexane molecule would give by inversion one (*R*)-iodohexane molecule, which could pair off with an unreacted (*S*)-iodohexane molecule causing racemisation.
In theory, reaction of only half the molecules could produce complete racemisation, but in fact as (S)-molecules are formed, they too can be re-substituted and go back to (R)-. Eventually there will be an equilibrated random mixture of (S)- and (R)-enantiomers of 2-iodohexane present.

SUMMARISE

Racemisation occurs by the bimolecular S_N2 mechanism in which iodide ions act as both the nucleophile and the leaving group. The rate of racemisation is faster than the rate of iodide ion exchange.

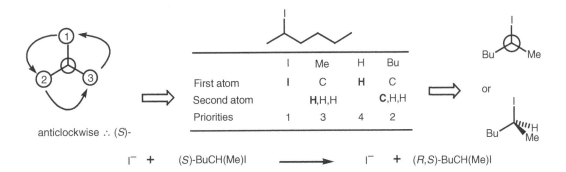

	I	Me	H	Bu
First atom	**I**	C	**H**	C
Second atom		**H**,H,H		**C**,H,H
Priorities	1	3	4	2

anticlockwise ∴ (S)-

I⁻ + (S)-BuCH(Me)I ⟶ I⁻ + (R,S)-BuCH(Me)I

I⁻ + RI ⟶ I⁻ + RI

HOMO: non-bonded pair of electrons, Nu⁻; LUMO: σ* of C–LG

S_N1
racemisation
(unimolecular)

rate-limiting
step
–LG⁻

symmetrical
intermediate

Nu⁻
to either
face

racemisation

S_N2
inversion
(bimolecular)

rate-limiting
step

achiral
transition state

rotate 180°

enantiomers

S_N2

When one molecule reacts   (S)-RI   —inversion→   (R)-RI   which 'pairs off' with another unreacted (S)-RI to form a racemic mixture

Mechanism is S_N2; rate of racemisation is twice as fast as the rate of inversion

3. Predict the structure of the epoxides from the reaction of NaOH (1 equivalent) with the 1,2-chloroalcohols **A** and **B**.

racemic
A

racemic
B

DRAW the STRUCTURES of A and B in three dimensions	Each 1,2-chloroalcohol has two chiral centres. *Newman projections are often useful for acyclic compounds. Don't forget to consider both chair conformers for **B**, and remember that larger groups prefer the more stable equatorial positions to avoid unfavourable axial–axial interactions across the ring.*
CONSIDER the REAGENTS and OUTLINE a MECHANISM	HO⁻ could act as a base (either to deprotonate the alcoholic OH or to induce elimination) or as a nucleophile in a direct substitution reaction. Using the alkoxide ion to make an epoxide by an internal S_N2 substitution will probably be the major reaction. This requires antiperiplanar $-O^-$ and $-Cl$. *As water and alcohols have similar acidity, there should be plenty of alkoxide present. The internal S_N2 to make an epoxide is favoured entropically and usually occurs in preference to β-elimination or displacement of Cl^- by external ^-OH, provided that the necessary antiperiplanar conformation can be achieved.*
DRAW out a possible ANSWER and predict the major PRODUCT(S)	Epoxide formation is possible from **A**, and from **B**. For **A**, antiperiplanar alkoxide and Cl is achieved by rotation about the central C–C. The resulting epoxide **C** has a plane of symmetry. For **B**, the less favourable diaxial conformation provides the correct stereochemistry; again the resulting epoxide **D** is symmetrical. *Both **C** and **D** are meso-structures, each having two chiral centres and a plane of symmetry.*

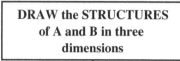

SUMMARISE	Treatment of 1,2-chloroalcohols **A** and **B** with NaOH gives epoxides **C** and **D** respectively. For efficient formation of epoxides the intermediate alkoxide $-O^-$ needs to be antiperiplanar to the chloride leaving group. The major conformer of **B** can also be involved in a ring-contraction process (see page 130).

A or

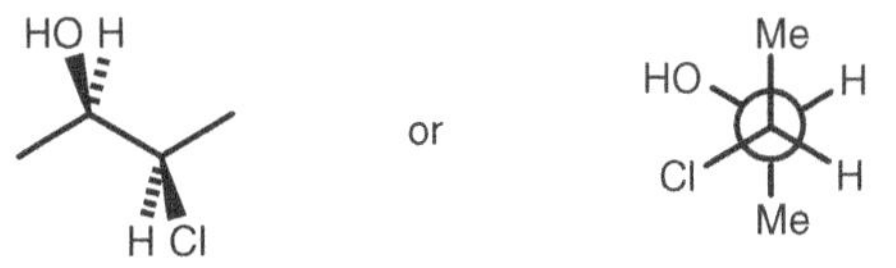

B

Cl$_{ax}$, OH$_{ax}$
(minor conformer)

ring flip

Cl$_{eq}$, OH$_{eq}$
(major conformer)

$$RCH(OH)CH(Cl)R \xrightarrow{NaOH} RHC\overset{O}{\underset{}{\diagup\diagdown}}CHR \ + \ NaCl \ + \ H_2O$$

S$_N$2 ring closure after deprotonation

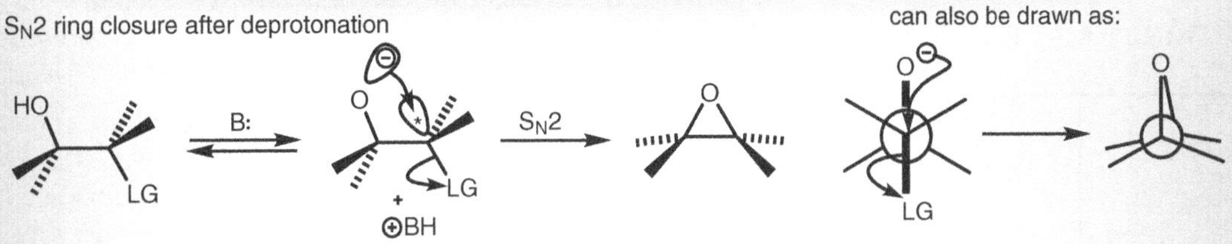

can also be drawn as:

B: S$_N$2

Epoxide ring closure requires the alkoxide and leaving group to be antiperiplanar
HOMO: non-bonded pair of electrons, $-O^-$; LUMO: σ^* of C–LG

A

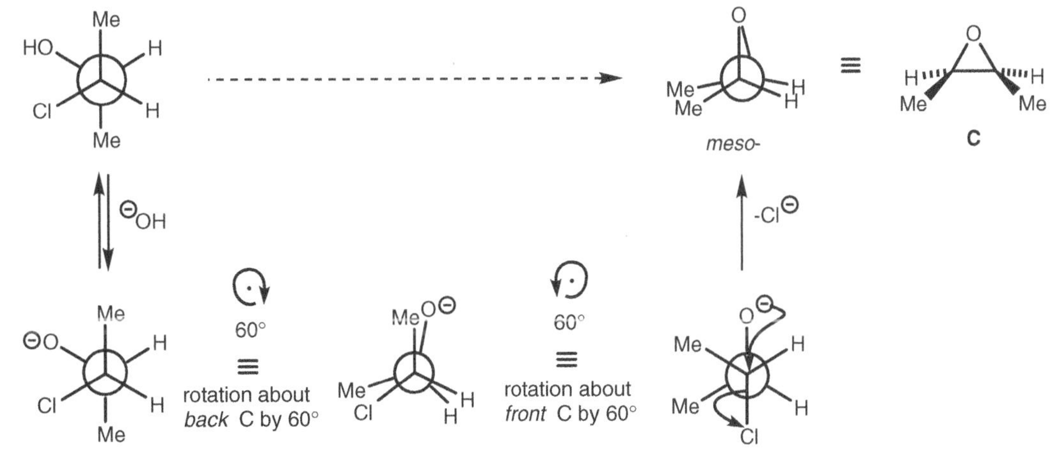

$\ominus$OH

60°
≡
rotation about
back C by 60°

60°
≡
rotation about
front C by 60°

-Cl$^\ominus$

meso-

C

B   The antiperiplanar substituents are axial, axial.

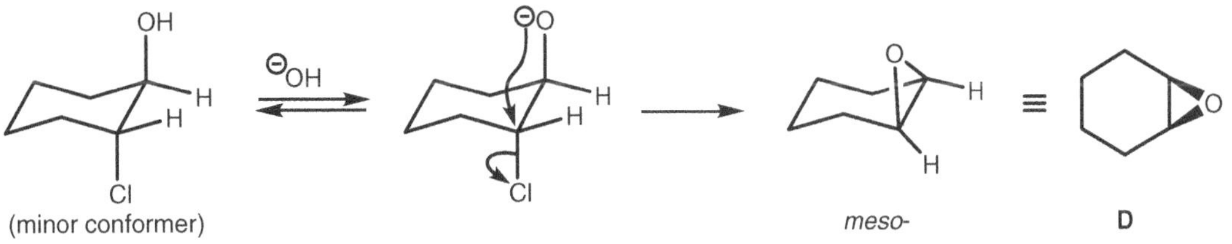

OH

$\ominus$OH

(minor conformer)

meso-

D

4. Deduce the structure of **B**, and explain the stereochemistry of the reaction.

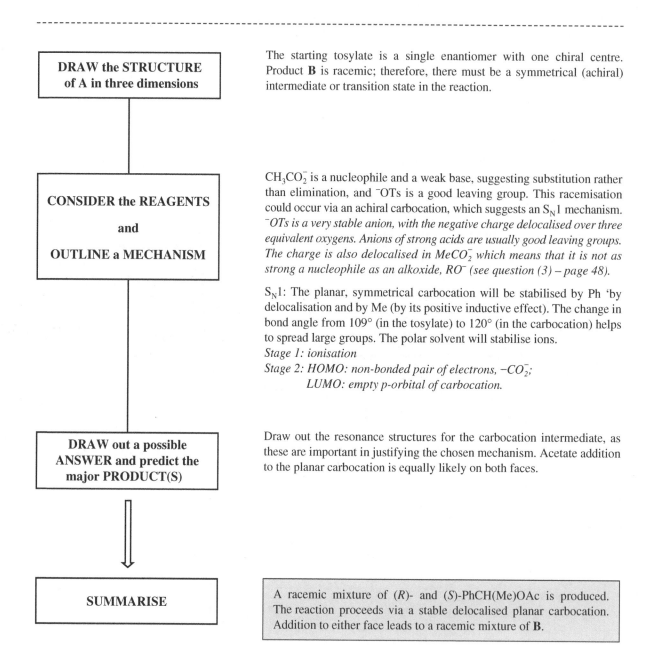

(R)-PhCH(Me)OTs $\xrightarrow[\text{in } CH_3CO_2H]{CH_3CO_2Na}$ $(\pm)$-**B** $(C_{10}H_{12}O_2)$

A

DRAW the STRUCTURE of A in three dimensions	The starting tosylate is a single enantiomer with one chiral centre. Product **B** is racemic; therefore, there must be a symmetrical (achiral) intermediate or transition state in the reaction.
CONSIDER the REAGENTS and OUTLINE a MECHANISM	$CH_3CO_2^-$ is a nucleophile and a weak base, suggesting substitution rather than elimination, and $^-$OTs is a good leaving group. This racemisation could occur via an achiral carbocation, which suggests an S_N1 mechanism. ^-OTs is a very stable anion, with the negative charge delocalised over three equivalent oxygens. Anions of strong acids are usually good leaving groups. The charge is also delocalised in $MeCO_2^-$ which means that it is not as strong a nucleophile as an alkoxide, RO^- (see question (3) – page 48).
	S_N1: The planar, symmetrical carbocation will be stabilised by Ph 'by delocalisation and by Me (by its positive inductive effect). The change in bond angle from 109° (in the tosylate) to 120° (in the carbocation) helps to spread large groups. The polar solvent will stabilise ions. *Stage 1: ionisation* *Stage 2: HOMO: non-bonded pair of electrons, $-CO_2^-$;* *LUMO: empty p-orbital of carbocation.*
DRAW out a possible ANSWER and predict the major PRODUCT(S)	Draw out the resonance structures for the carbocation intermediate, as these are important in justifying the chosen mechanism. Acetate addition to the planar carbocation is equally likely on both faces.
SUMMARISE	A racemic mixture of (R)- and (S)-PhCH(Me)OAc is produced. The reaction proceeds via a stable delocalised planar carbocation. Addition to either face leads to a racemic mixture of **B**.

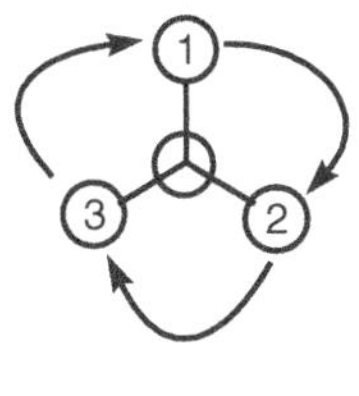
1
2
3

(*R*)- ∴ clockwise

	Ph	H	CH_3	OTs
First atom	C	**H**	C	**O**
Second atom	**C**,C,C		**H**,H,H	
Priorities	2	4	3	1

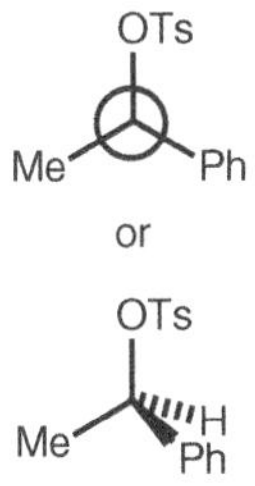
OTs
Me
Ph
or
OTs
Me
H
Ph

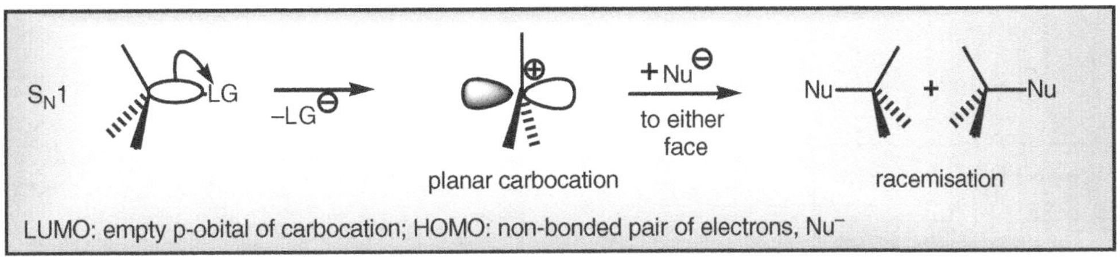
S_N1
LG
–LG⊖
planar carbocation
+Nu⊖
to either
face
Nu
+
Nu
racemisation
LUMO: empty p-obital of carbocation; HOMO: non-bonded pair of electrons, Nu⁻

Major reaction is S_N1.
AcO^- is a weak base, so little competing E1 elimination to $PhCH{=}CH_2$ is expected.

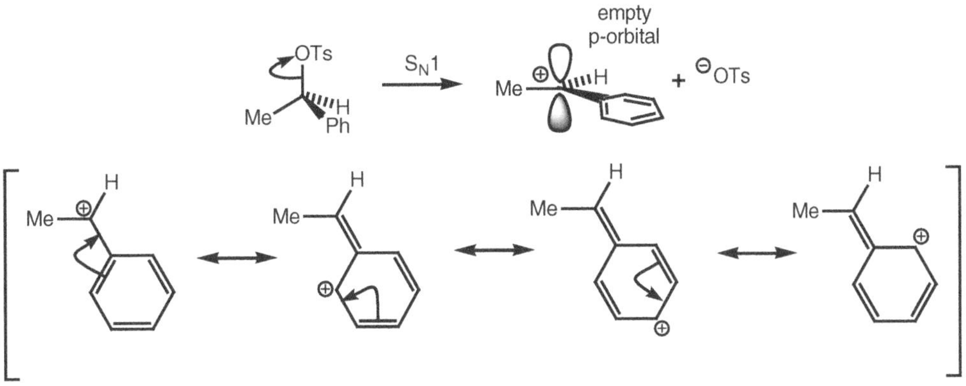
OTs
Me
H
Ph
SN1
empty
p-orbital
Me
H
+
⊖OTs
Me
H
Me
H
Me
H
Me
H

Then,

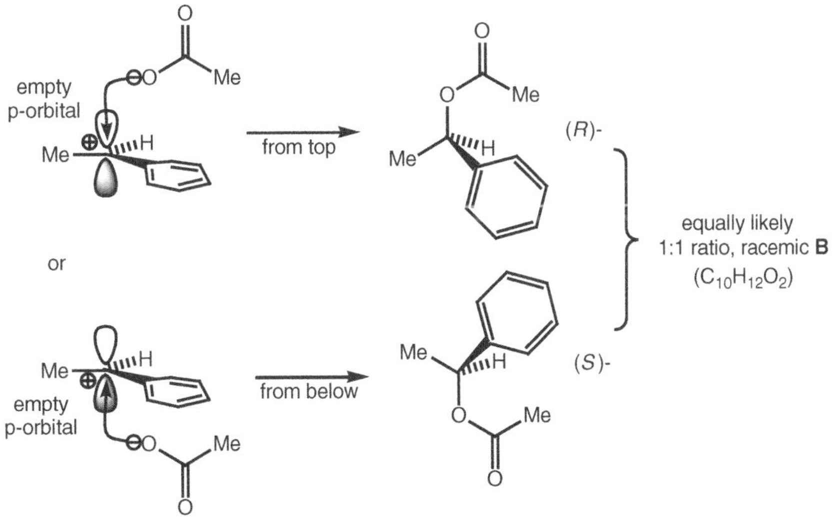
O
O
Me
empty
p-orbital
Me
H
from top
O
O
Me
Me
H
(R)-
equally likely
1:1 ratio, racemic B
(C10H12O2)
or
Me
H
empty
p-orbital
O
Me
O
from below
Me
H
(S)-
O
Me
O

5. Rationalise the following reaction.

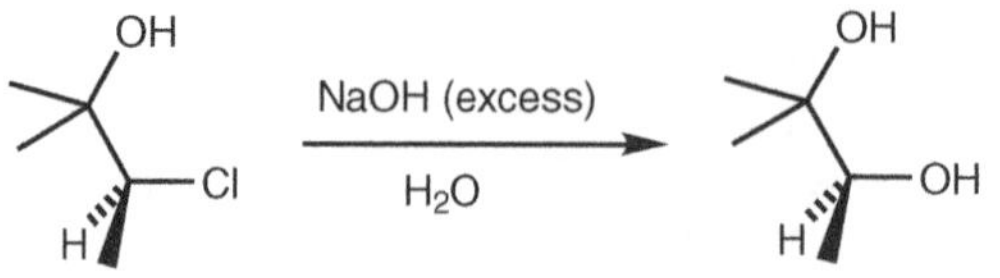

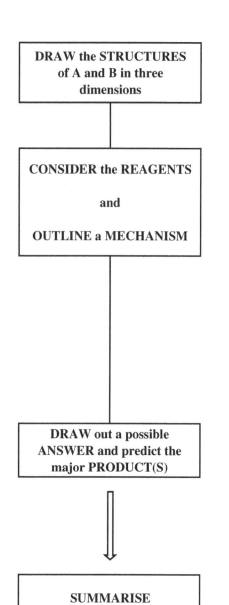

(S)-A   NaOH (excess) / H₂O →   (S)-B

DRAW the STRUCTURES of A and B in three dimensions	Both **A** and **B** have a single chiral centre. Inspection of the structures shows that OH takes the place of Cl with retention of stereochemistry.
CONSIDER the REAGENTS and OUTLINE a MECHANISM	HO⁻ is a nucleophile and a base. Because of the retention of configuration in **B**, this reaction is neither a direct substitution reaction by a single S_N2 reaction (which would go with inversion) nor an S_N1 reaction (for which racemisation is expected). It could go by two consecutive inversions, which would give overall retention via an intermediate. The intermediate could be an epoxide. HO⁻ can act as a base to deprotonate the alcohol to give an alkoxide, which acts as an internal nucleophile in the first S_N2 reaction to give the epoxide. Ring-opening by HO⁻ in the second S_N2 reaction leads to the diol with overall retention of configuration. *This is a neighbouring group effect (see page 42). The hydroxyl group provides the internal nucleophile in the first stage and is re-formed as the leaving group in the second stage. First stage: HOMO: −O⁻ non-bonded pair; LUMO: σ* of C−Cl. Second stage: HOMO: ⁻OH non-bonded pair; LUMO: σ* of C−O.*
DRAW out a possible ANSWER and predict the major PRODUCT(S)	For epoxide formation, the C−O⁻ bond and the C−Cl bond must be antiperiplanar so that there is a good overlap between the σ* of the C−Cl and a non-bonded pair of electrons on −O⁻. The ⁻OH then attacks the less hindered of the two epoxide carbons by an S_N2 mechanism. *As in question (2) (on page 46), the S_N2 mechanisms are drawn with the bonds being broken and formed in the plane of the paper so that the stereochemical changes are easier to follow.*
SUMMARISE	Deprotonation of the ROH with ⁻OH is rapid and reversible. Intramolecular S_N2 displacement of chloride ion gives an unsymmetrical epoxide with inversion of configuration. Ring opening with ⁻OH at the less hindered carbon of the epoxide gives the diol **B**, again with inversion of configuration. Both reactions occur with inversion leading to overall retention of configuration.

(see page 42)
on page 46

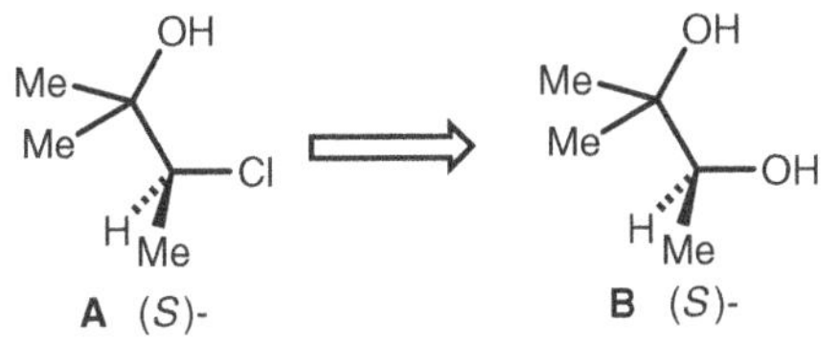

∴ Retention of configuration observed.
Could be *two* S$_N$2 reactions (two inversions are equivalent to one retention).
The intermediate could be the epoxide.

Not S$_N$1: racemisation expected
Not S$_N$2: inversion, (*R*)-configuration would be expected

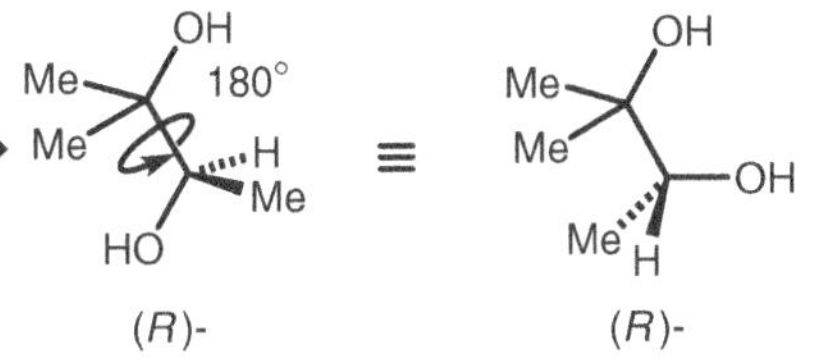

$$RCH(OH)CH(Cl)R \xrightarrow{NaOH} RHC\overset{O}{-}CHR + NaCl + H_2O$$

A double S$_N$2 reaction

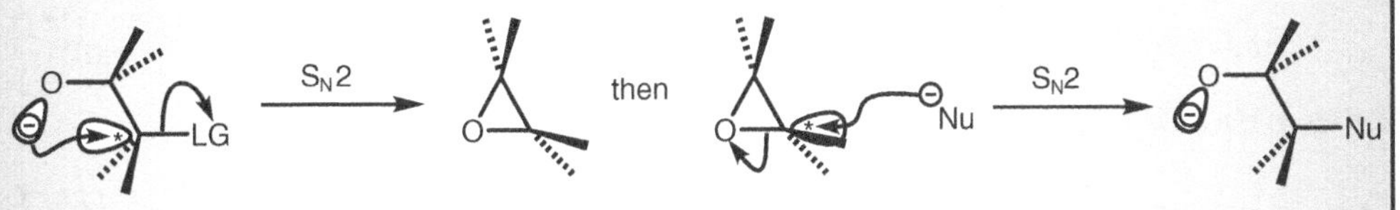

HOMO: non-bonded pair of electrons, –O⁻; LUMO: σ* of C–LG

Closing of epoxide requires: antiperiplanar –O⁻ and LG
Opening of epoxide requires: linear Nu–C–O

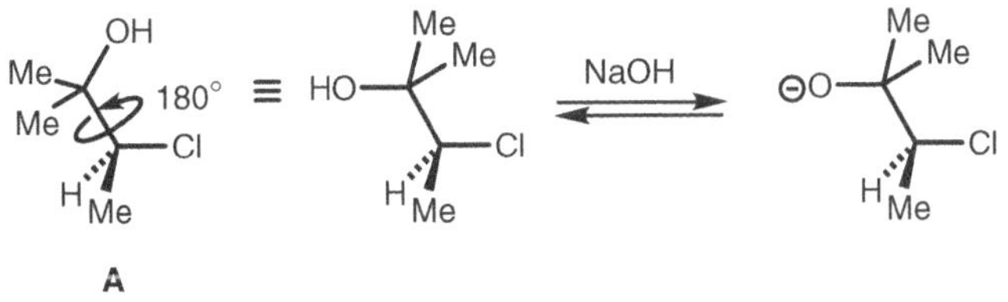

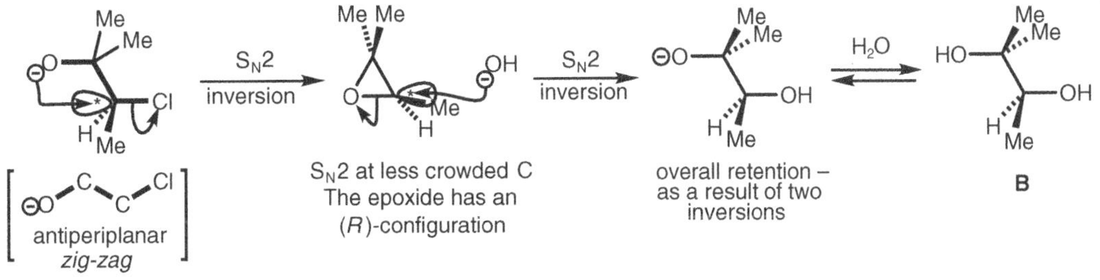

antiperiplanar *zig-zag*

S$_N$2 at less crowded C
The epoxide has an (*R*)-configuration

overall retention – as a result of two inversions

6. Suggest an explanation for the observed behaviour of the chiral tosylate **A** with NaOAc in AcOH.

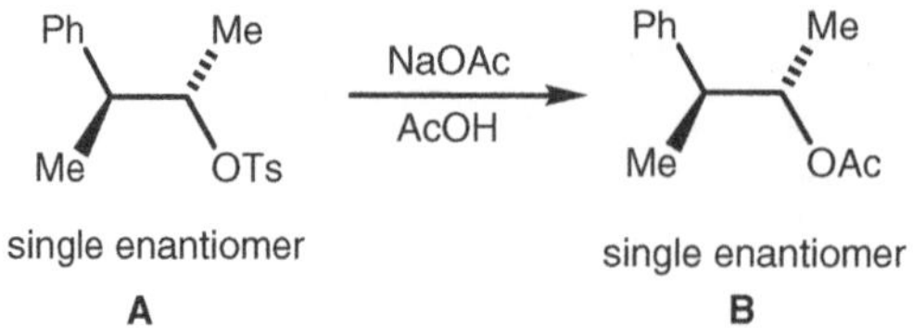

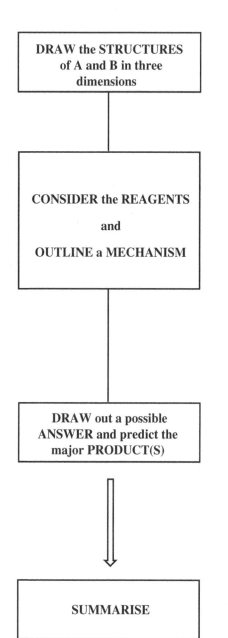

single enantiomer
A

single enantiomer
B

DRAW the STRUCTURES of A and B in three dimensions	Tosylate **A** is converted into acetate **B**. Each has two chiral centres and inspection of the structures shows that the reaction goes with overall **retention** of configuration. *Newman projections are often useful for acyclic compounds with two chiral centres.*
CONSIDER the REAGENTS **and** **OUTLINE a MECHANISM**	The overall retention of configuration suggests a double inversion. Conversion of **A** into **B** cannot involve a symmetrical intermediate (or transition state) because the product is chiral. Two consecutive S$_N$2 reactions on the carbon carrying OTs could use the π electrons of the Ph as the *neighbouring group*, the intermediate being a cyclic cation. *Stage 1: HOMO: π of Ph; LUMO: σ^* of C–OTs.* *Stage 2: HOMO: non-bonded pair, $-CO_2^-$; LUMO: σ^* of C–C.*
DRAW out a possible ANSWER and predict the major PRODUCT(S)	The electron-rich Ph group can act as the first internal nucleophile by assisting in the displacement of OTs, making the bridged, resonance stabilised carbocation **C**. **C** is chiral; attack at either carbon gives chiral acetate **B**. The phenyl ring acts as a neighbouring group in this reaction through participation of its π-electrons, and remains unchanged at the end of the reaction. *Cation **C** is chiral but it still possesses a two-fold rotation axis (C_2). Addition to either the front or back carbon atom leads to the same acetate **B**.*
SUMMARISE	The Ph group participates in the elimination of TsO$^-$ to give the resonance stabilised cyclic cationic intermediate **C**. S$_N$2 ring-opening by AcO$^-$ leads to a single enantiomer of **B**.

A A B

Retention
NO symmetrical stage allowed
Not S_N1, not a single S_N2

S_N2, twice

Neighbouring group (Nu) is Ph group
Stage a: HOMO: non-bonded pair of electrons, Nu⁻; LUMO: σ^* of C–LG
Stage b: HOMO: non-bonded pair of electrons, Y⁻ ; LUMO: σ^* of C–Nu⁺

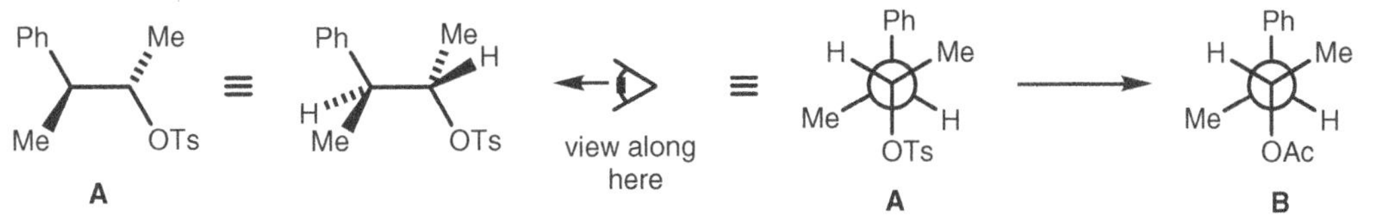

For **A**:

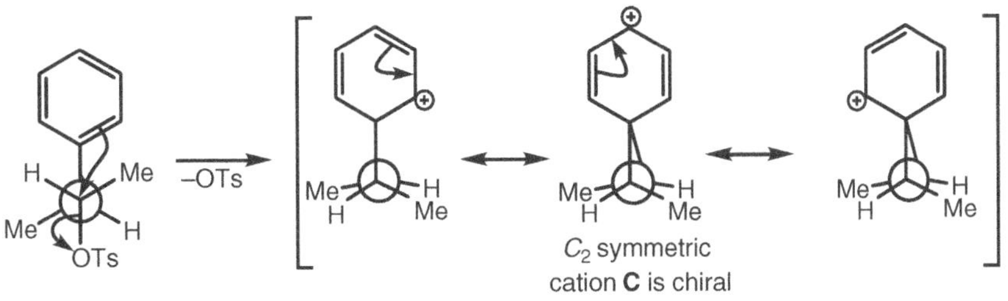

then *either*

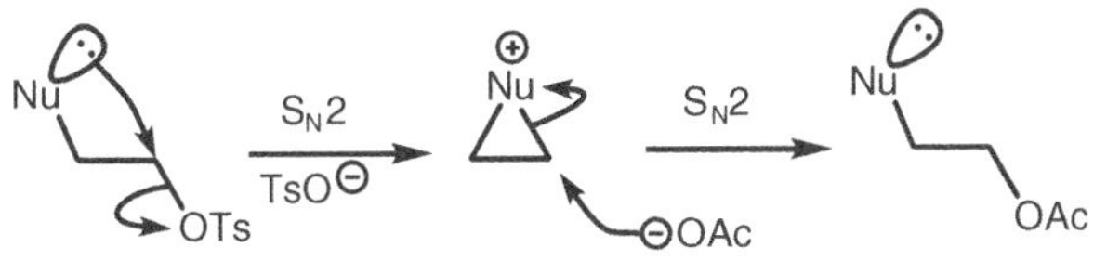

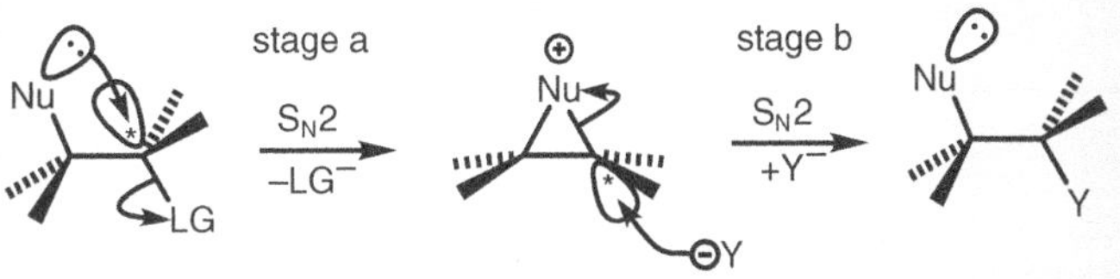

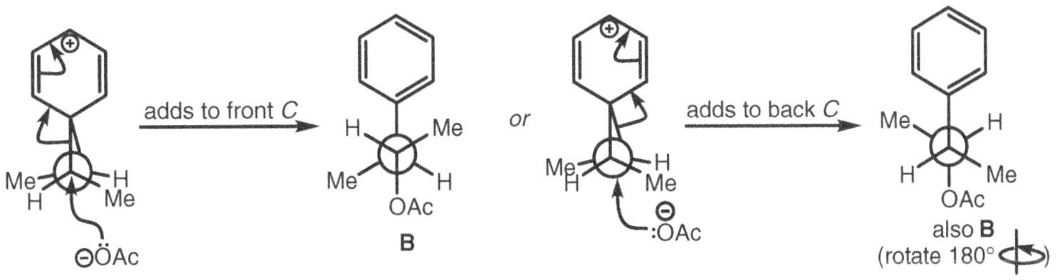

7. Deduce the full structures and stereochemistry for **A**, **B** and **C** in the following sequence, in which the reaction of **A** to **B** was found to show bimolecular kinetics.

A; PhCH(Me)Cl $\xrightarrow{\text{NaN}_3}$ **B** $\xrightarrow{\text{H}_2/\text{Pd-C}}$ **C**; (*S*)-PhCH(Me)NH$_2$

single enantiomer

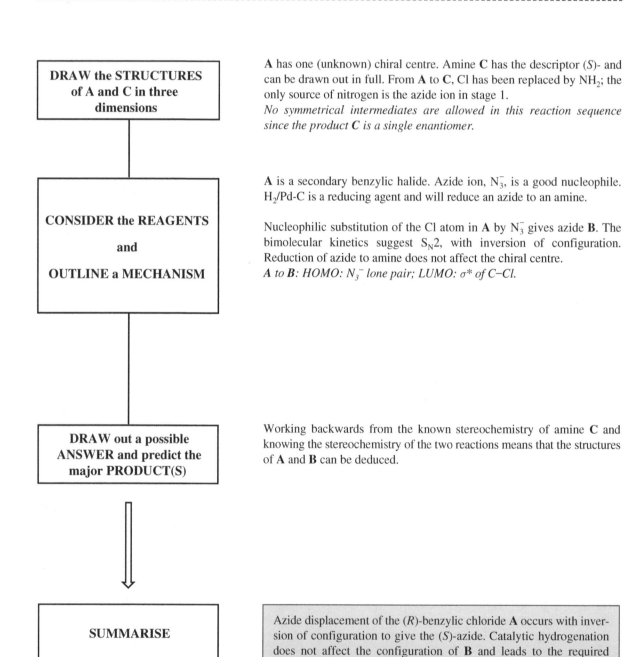

DRAW the STRUCTURES of A and C in three dimensions

A has one (unknown) chiral centre. Amine **C** has the descriptor (*S*)- and can be drawn out in full. From **A** to **C**, Cl has been replaced by NH$_2$; the only source of nitrogen is the azide ion in stage 1.
*No symmetrical intermediates are allowed in this reaction sequence since the product **C** is a single enantiomer.*

CONSIDER the REAGENTS and OUTLINE a MECHANISM

A is a secondary benzylic halide. Azide ion, N$_3^-$, is a good nucleophile. H$_2$/Pd-C is a reducing agent and will reduce an azide to an amine.

Nucleophilic substitution of the Cl atom in **A** by N$_3^-$ gives azide **B**. The bimolecular kinetics suggest S$_N$2, with inversion of configuration. Reduction of azide to amine does not affect the chiral centre.
A to B: HOMO: N$_3^-$ lone pair; LUMO: σ* of C−Cl.

DRAW out a possible ANSWER and predict the major PRODUCT(S)

Working backwards from the known stereochemistry of amine **C** and knowing the stereochemistry of the two reactions means that the structures of **A** and **B** can be deduced.

SUMMARISE

Azide displacement of the (*R*)-benzylic chloride **A** occurs with inversion of configuration to give the (*S*)-azide. Catalytic hydrogenation does not affect the configuration of **B** and leads to the required (*S*)-amine. Overall, there is an inversion of configuration from **A** to **C**.

A: Ph—C(—Me)(H)(Cl)   (*R*)- or (*S*)-

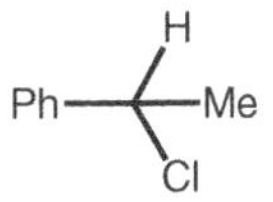

unknown configuration

B: unknown

	NH$_2$	Ph	Me	H
First atom	**N**	C	C	**H**
Second atom		**C**,C,C	**H**,H,H	
Priorities	1	2	3	4

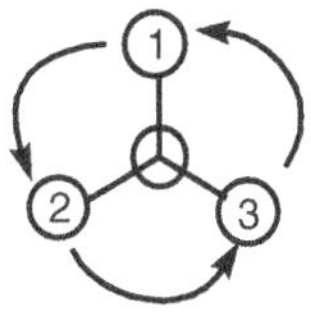

C: (*S*)-PhCH(Me)NH$_2$

anticlockwise for (*S*)-

anticlockwise

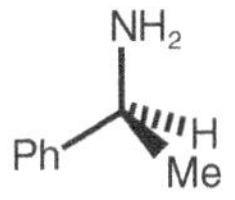

∴(*S*)-

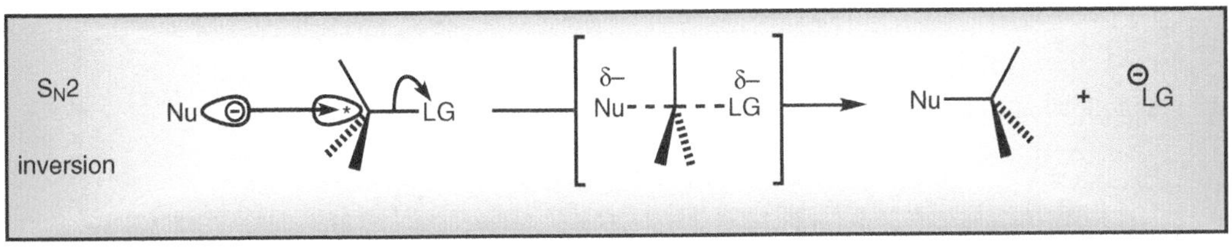

RCl $\xrightarrow[\substack{S_N 2 \\ \text{inversion}}]{\text{NaN}_3}$ RN$_3$ $\xrightarrow[\text{retention}]{\text{H}_2/\text{Pd-C}}$ RNH$_2$

S$_N$2

inversion

$$\text{Nu}^{\ominus} \cdots \overset{*}{\underset{}{C}} \text{—LG} \longrightarrow \left[\overset{\delta-}{\text{Nu}} \text{---} \overset{\delta-}{\text{LG}} \right] \longrightarrow \text{Nu—}C + {}^{\ominus}\text{LG}$$

Working backwards:

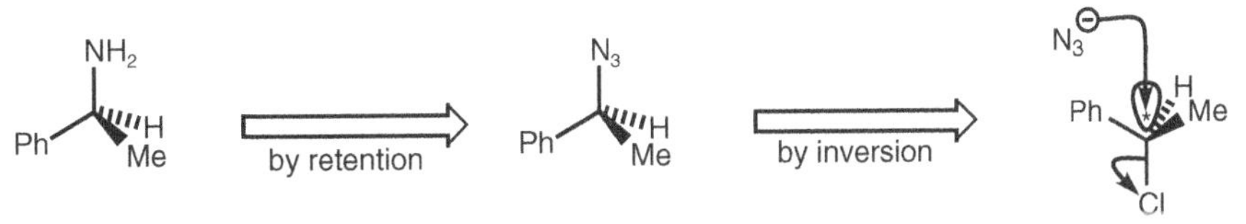

C		**B**		**A**
Ph—C(NH$_2$)(H)(Me)	by retention	Ph—C(N$_3$)(H)(Me)	by inversion	N$_3^{\ominus}$ ... Ph—C(H)(Me)—Cl
priorities		priorities		priorities
NH$_2$ > Ph > Me > H		N$_3$ > Ph > Me > H		Cl > Ph > Me > H

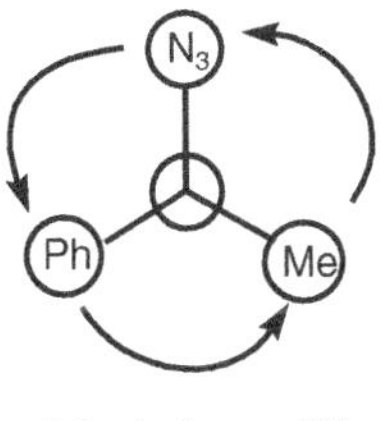

anticlockwise, ∴ (*S*)-

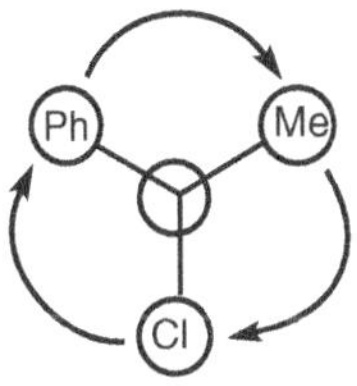

clockwise, ∴ (*R*)-

Additional problems for Chapter 2

1. Predict whether retention or inversion of configuration occurs in the following reactions.

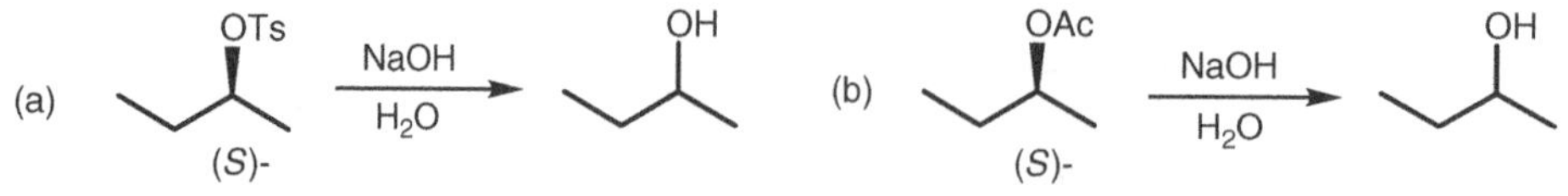

(a) OTs — NaOH / H_2O → OH   (S)-

(b) OAc — NaOH / H_2O → OH   (S)-

2. Suggest reagents for the conversion of (R)-2-hexanol into its other enantiomeric form.

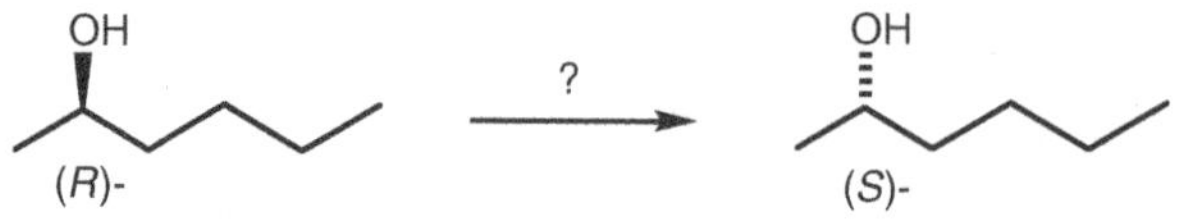

OH (R)- —?→ OH (S)-

3. Account for the stereochemistry of the following reaction.

Me, H_2N, CO_2H (S)- — NaNO₂ / HBr → Me, Br, CO_2H (S)-

4. Explain the following observation.

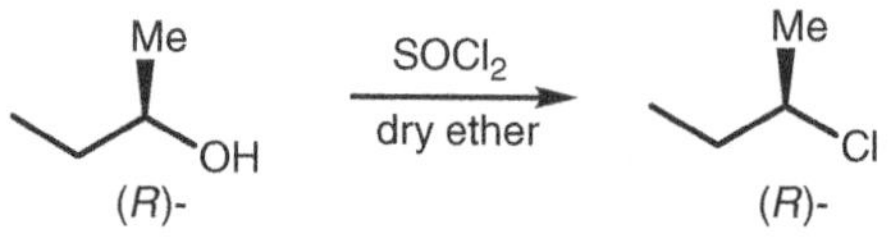

Me, OH (R)- — SOCl₂ / dry ether → Me, Cl (R)-

5. Explain why the displacement of the tertiary bromide with NaOEt occurs via a direct S_N2 mechanism rather than an S_N1 mechanism.

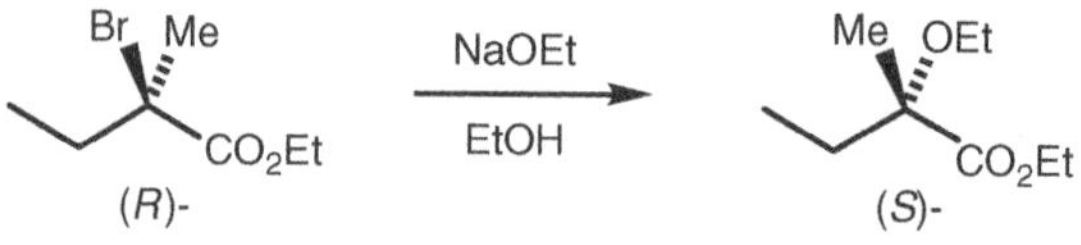

Br, Me, CO_2Et (R)- — NaOEt / EtOH → Me, OEt, CO_2Et (S)-

6. Suggest and explain the mechanism for the following interconversion.

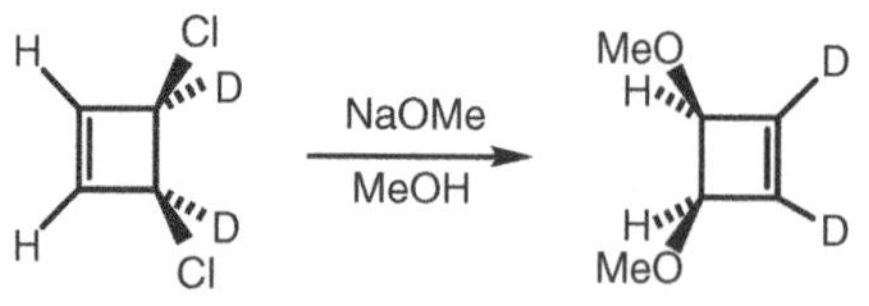

H, Cl, D, H, Cl, D — NaOMe / MeOH → MeO, H, D, H, MeO, D

Answers: 1. (a) Inversion, S_N2 displacement, ⁻OTs good leaving group; (b) retention; hydrolysis of ester by C=O addition and elimination; 2. A possible solution: convert alcohol to tosylate (TsCl in pyridine – see question (1)), then displace with KOAc to give the corresponding acetate (with inversion of configuration – S_N2). Hydrolysis of the intermediate acetate (with NaOH/H_2O – see question (1b) above) would give the required (S)-2-hexanol; 3. Nitrous acid (HNO₂) is immediately formed by addition of NaNO₂ and HCl. This converts the amino group (NH₂) into a diazonium group (N₂⁺). S_N2 displacement (with inversion of configuration) and loss of N₂ by reaction with the neighbouring carboxylic acid gives an intermediate α-lactone. S_N2 ring opening at the C–O bond and inversion of configuration with a bromide ion gives the required bromide. Overall retention of configuration due to two consecutive inversions; 4. S_Ni mechanism (see page 43); 5. S_N1 displacement is not favoured due to adjacent electron withdrawing carbonyl substituent destabilising carbocation formation. This lowers the relative S_N1 reaction rate, and the S_N2 reaction is preferred; 6. Two consecutive S_N2' displacements – addition of the incoming nucleophile (⁻OMe) must occur on the same face as the leaving group (Cl) (see page 42).

3 β-Elimination Reactions to form Alkenes

The aim of this chapter is to develop an understanding of the stereochemical features of various mechanisms for the formation of alkenes by β-elimination.

At the end of this chapter, you should be able to:

- Draw and compare the mechanisms for E1, E2, E1$_{cb}$ and cyclic polar eliminations, the differences between these and their competition with substitution.
- Predict and explain the stereochemical consequences of these reactions.
- Explain the influence of adjacent groups on the mechanisms of these reactions.

Background: β-elimination reactions

There are four major types of polar elimination reactions in which a proton and a leaving group, H–LG, are lost to form an alkene. Three of them E1, E2 and E1$_{cb}$,[3] describe a continuous spectrum of possible mechanisms (in the way that S$_N$1 and S$_N$2 do) in which the leaving group LG$^-$ leaves first, H and LG leave together, or H$^+$ leaves first. The remaining cyclic eliminations will be considered separately.

	E1	E2	E1$_{cb}$	cyclic transition state
Rate determining step	loss of leaving group, LG$^-$	loss of H and LG	loss of H$^+$	loss of H and LG
Intermediate	carbocation	none (concerted acyclic transition state)	carbanion	none (concerted cyclic transition state)
Examples	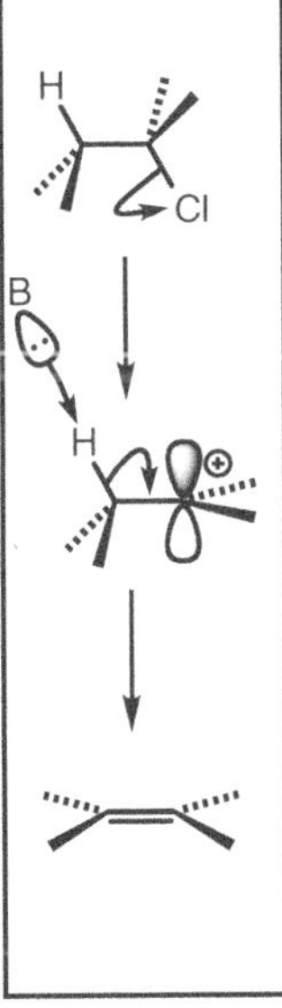	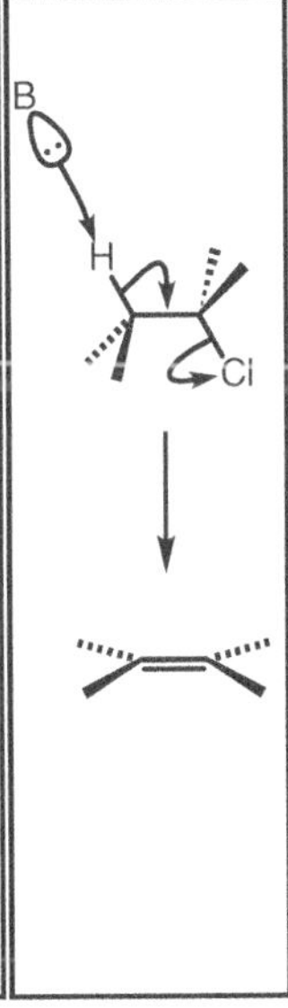	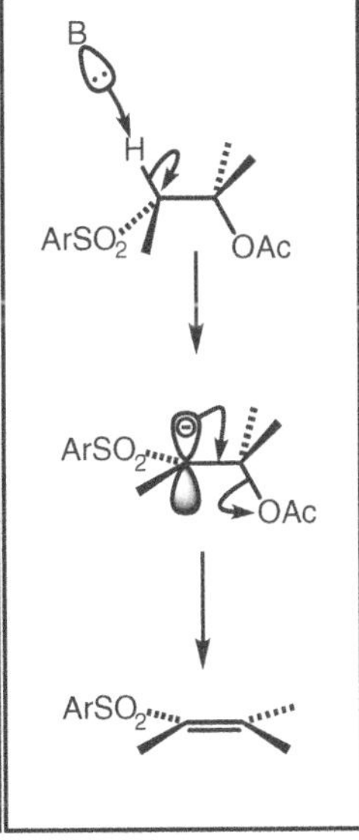	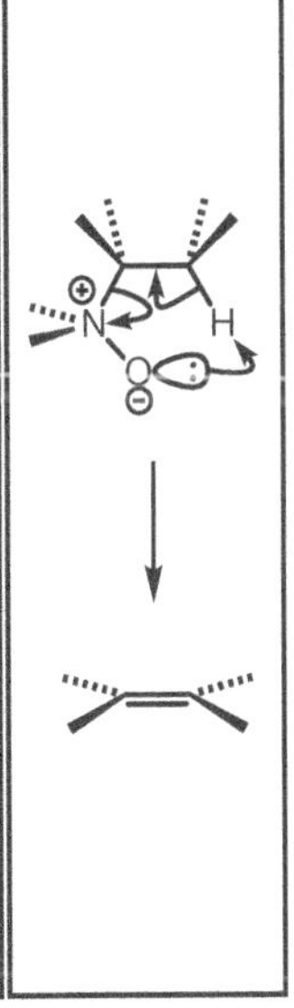

---

[3] In E1$_{cb}$, *cb* stands for *conjugate base*, which is the intermediate anion.

E1 eliminations

Loss of the leaving group LG gives a planar carbocation intermediate. Sideways overlap of the σ C–H orbital (HOMO) with the LUMO of the cation results in the loss of H^+ to make the new π-bond in the second stage.

Since both E1 and S_N1 have the same intermediate carbocation, there will be competition between these, depending on whether the reagent is acting more as a base (E1) or as a nucleophile (S_N1).

E2 eliminations

Sideways overlap of the σ C–H (HOMO) with the σ* of the polar C–LG bond (LUMO) leads to *anti*-elimination of H–LG in a single stage.

H and LG need to be antiperiplanar for efficient E2 elimination to take place. A rare, much slower *syn*-elimination via a synperiplanar transition state can occur if an antiperiplanar conformation is impossible, but the compound must have a suitable fixed synperiplanar conformation.

E2 and S_N2 mechansims often compete; again elimination requires a base, whereas substitution requires a nucleophile.

E1$_{cb}$ eliminations

Loss of the proton first gives a planar (or rapidly inverting) carbanion, which is usually stabilised by a strongly electron-withdrawing group (EWG). Overlap of the anion HOMO with the LUMO (σ*) of the C–LG bond results in the loss of LG^- to make the new π-bond in the second stage.

(EWG = electron withdrawing group)

carbanion intermediate

The stereoselectivity of the reaction depends on the lifetime (and thus the stability) of the carbanion intermediate. If it is short, then more *anti*-elimination is expected as the mechanism tends towards E2. If the carbanion is stable, and C–C bond rotation is possible, then more of the more stable alkene is expected.

Choosing between E1, E2 and E1$_{cb}$

One way to approach the problem of assigning a mechanism to a given elimination is to assume that it will be E2 unless there are good reasons to think otherwise. For E1, look for a weak base, a good leaving group and a stabilised carbocation intermediate; for E1$_{cb}$, a strong base and a stabilised carbanion. If neither intermediate looks particularly favourable, stay with E2 elimination.

Cyclic eliminations

In these single-step processes, an intramolecular basic group (usually carrying a non-bonded pair of electrons) starts off the cyclic mechanism by attacking a β-proton. In acyclic compounds, the planar transition state required can easily be achieved using a synperiplanar conformation.

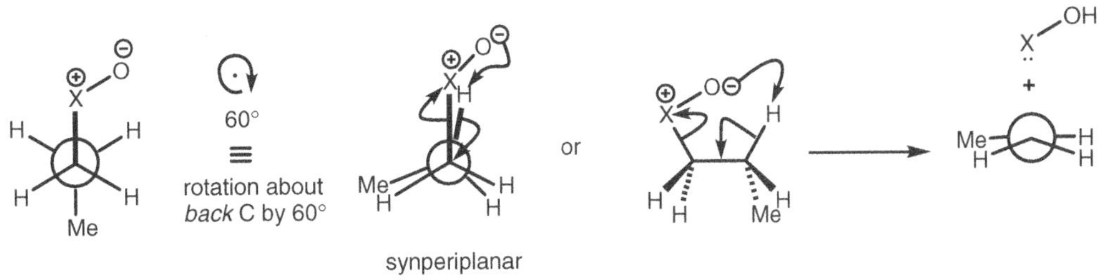

where $X^+\text{-}O^-$ can be $R_2N^+\text{-}O^-$, $RS^+\text{-}O^-$, $RSe^+\text{-}O^-$ and many more.

In cyclohexyl systems coplanarity can be achieved by a *syn-* axial–equatorial pair of adjacent substituents in a half-chair conformation. For example:

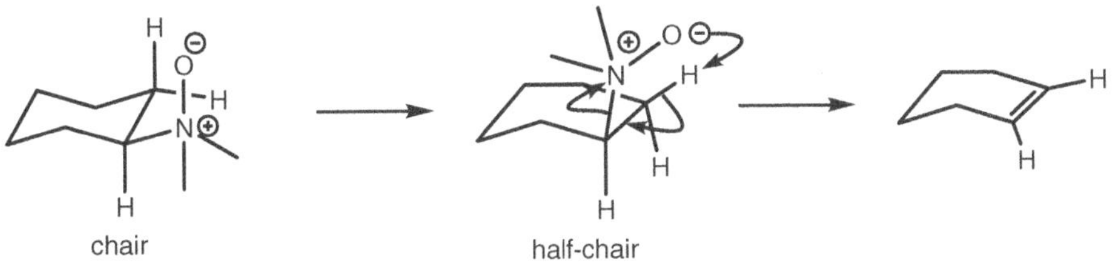

or, using Newman projections:

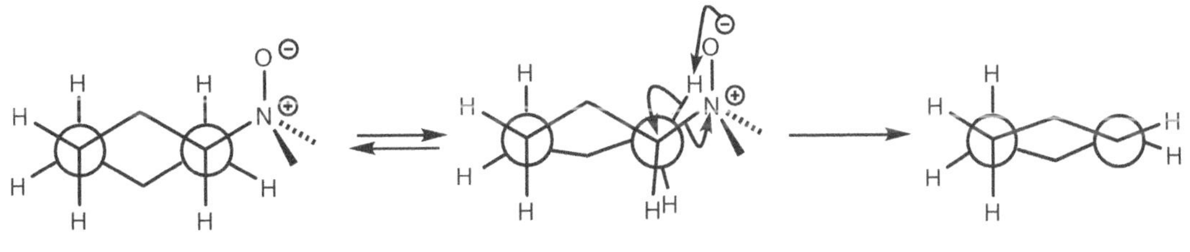

One end of the chair is flattened to achieve the axial–equatorial coplanarity of the half chair – this is quite easy to achieve – try it by using molecular models. Making an equatorial–equatorial pair coplanar is much harder because it involves greater puckering of the ring, forcing the axial substituents closer together.

These reactions all involve six electrons in a cyclic transition state, and can also be treated as *pericyclic group transfer reactions*, for which all the bond-forming and bond-breaking takes place in a single, planar and cyclic transition state. For further discussions, see Ian Fleming, *Pericyclic Reactions*, Oxford Chemistry Primer No 67, OUP, 1999.

1. Explain the different ratios of products in these reactions of 2-bromo, 2-methylbutane.

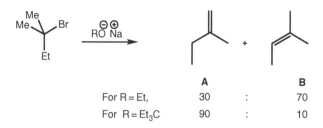

		A		B
For R = Et,		30	:	70
For R = Et$_3$C		90	:	10

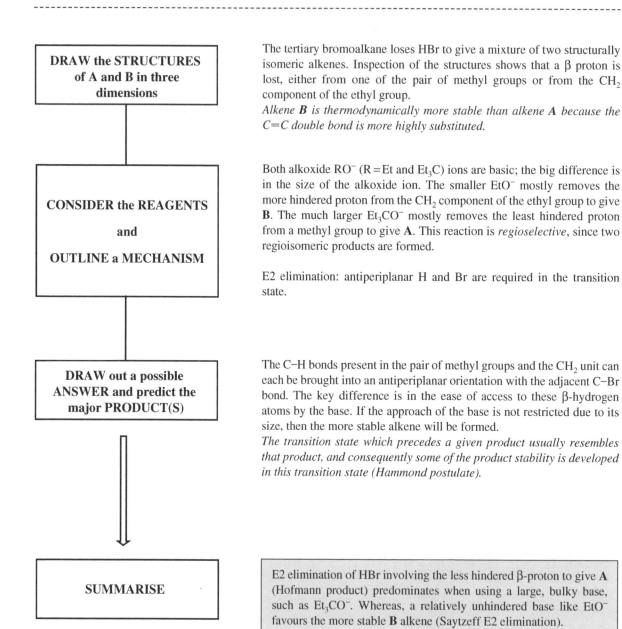

DRAW the STRUCTURES of A and B in three dimensions

The tertiary bromoalkane loses HBr to give a mixture of two structurally isomeric alkenes. Inspection of the structures shows that a β proton is lost, either from one of the pair of methyl groups or from the CH$_2$ component of the ethyl group.

*Alkene **B** is thermodynamically more stable than alkene **A** because the C=C double bond is more highly substituted.*

CONSIDER the REAGENTS and OUTLINE a MECHANISM

Both alkoxide RO$^-$ (R = Et and Et$_3$C) ions are basic; the big difference is in the size of the alkoxide ion. The smaller EtO$^-$ mostly removes the more hindered proton from the CH$_2$ component of the ethyl group to give **B**. The much larger Et$_3$CO$^-$ mostly removes the least hindered proton from a methyl group to give **A**. This reaction is *regioselective*, since two regioisomeric products are formed.

E2 elimination: antiperiplanar H and Br are required in the transition state.

DRAW out a possible ANSWER and predict the major PRODUCT(S)

The C—H bonds present in the pair of methyl groups and the CH$_2$ unit can each be brought into an antiperiplanar orientation with the adjacent C—Br bond. The key difference is in the ease of access to these β-hydrogen atoms by the base. If the approach of the base is not restricted due to its size, then the more stable alkene will be formed.

The transition state which precedes a given product usually resembles that product, and consequently some of the product stability is developed in this transition state (Hammond postulate).

SUMMARISE

E2 elimination of HBr involving the less hindered β-proton to give **A** (Hofmann product) predominates when using a large, bulky base, such as Et$_3$CO$^-$. Whereas, a relatively unhindered base like EtO$^-$ favours the more stable **B** alkene (Saytzeff E2 elimination).

product
stability

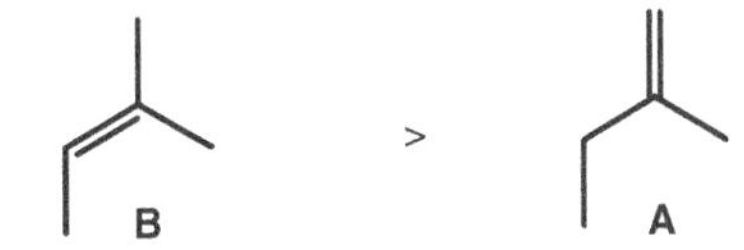

B   >   A

reagent
size   (CH₃CH₂)₃CO⊖   >   CH₃CH₂O⊖

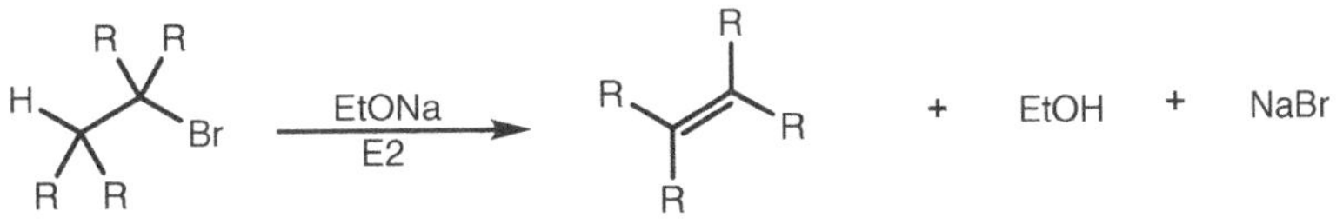

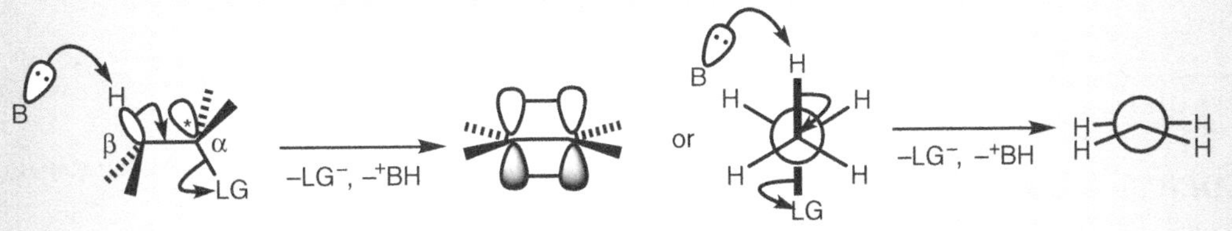

$(CH_3CH_2)_3CO^{\ominus}$ > $CH_3CH_2O^{\ominus}$

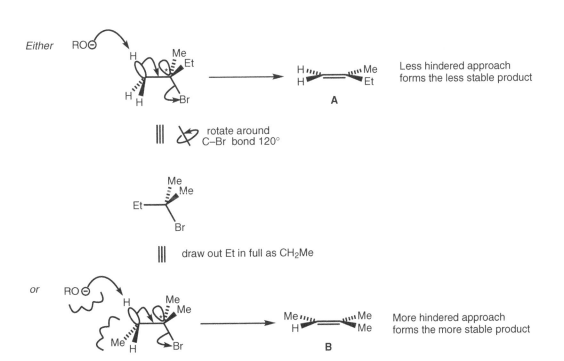

EtO⊖ is a smaller base, both approaches are easy, so the more stable alkene **B** predominates.

Et₃CO⊖ is a larger base, approach to form **B** is sterically hindered, so **A** predominates.

2. Explain the following results.

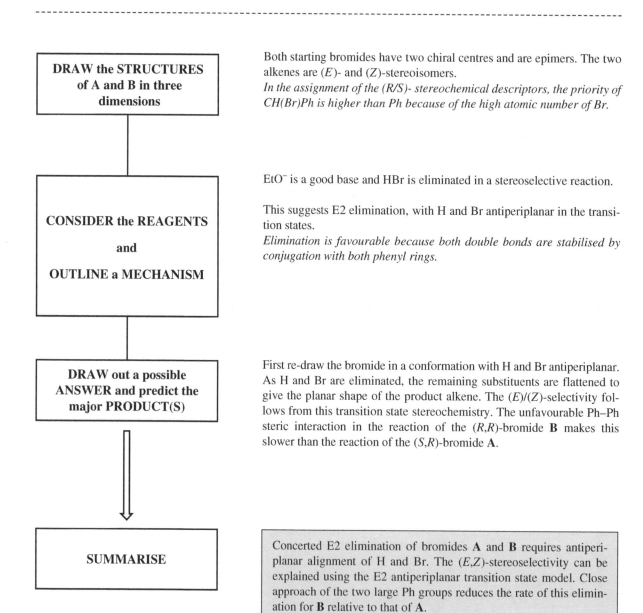

| DRAW the STRUCTURES of A and B in three dimensions | Both starting bromides have two chiral centres and are epimers. The two alkenes are (*E*)- and (*Z*)-stereoisomers.
In the assignment of the (R/S)- stereochemical descriptors, the priority of CH(Br)Ph is higher than Ph because of the high atomic number of Br. |

| CONSIDER the REAGENTS and OUTLINE a MECHANISM | EtO⁻ is a good base and HBr is eliminated in a stereoselective reaction.

This suggests E2 elimination, with H and Br antiperiplanar in the transition states.
Elimination is favourable because both double bonds are stabilised by conjugation with both phenyl rings. |

| DRAW out a possible ANSWER and predict the major PRODUCT(S) | First re-draw the bromide in a conformation with H and Br antiperiplanar. As H and Br are eliminated, the remaining substituents are flattened to give the planar shape of the product alkene. The (*E*)/(*Z*)-selectivity follows from this transition state stereochemistry. The unfavourable Ph–Ph steric interaction in the reaction of the (*R,R*)-bromide **B** makes this slower than the reaction of the (*S,R*)-bromide **A**. |

| SUMMARISE | Concerted E2 elimination of bromides **A** and **B** requires antiperiplanar alignment of H and Br. The (*E,Z*)-stereoselectivity can be explained using the E2 antiperiplanar transition state model. Close approach of the two large Ph groups reduces the rate of this elimination for **B** relative to that of **A**. |

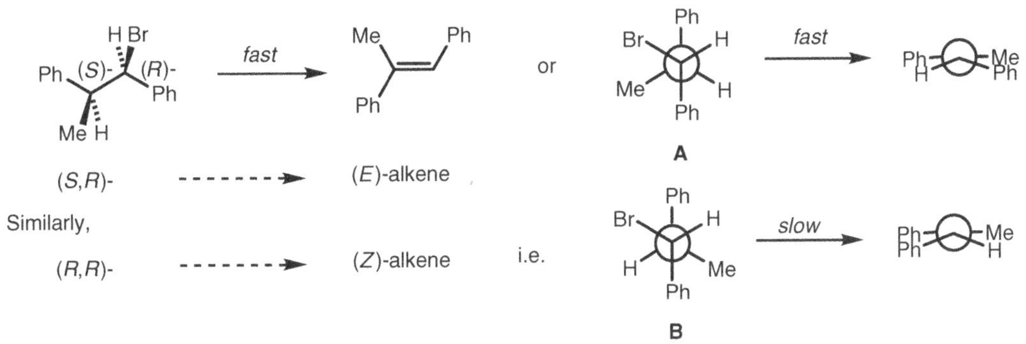

RCH₂CH₂Br $\xrightarrow[\text{E2}]{\text{NaOEt}}$ RCH=CH₂ + EtOH + NaBr

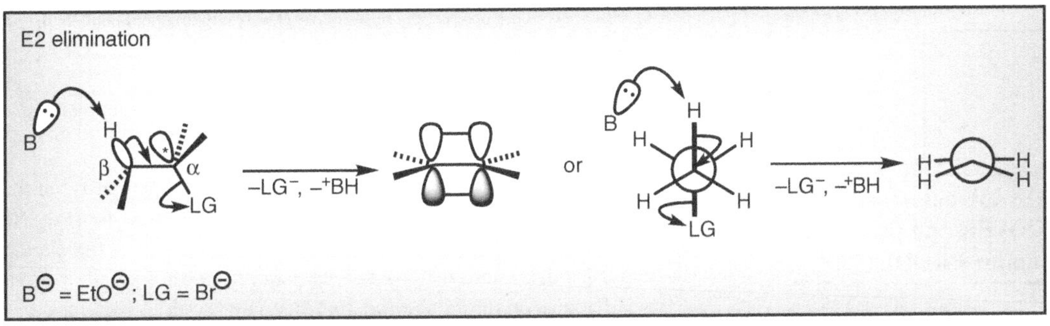

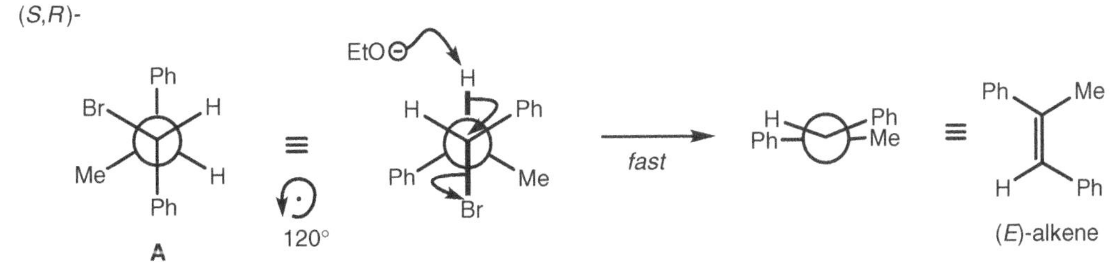

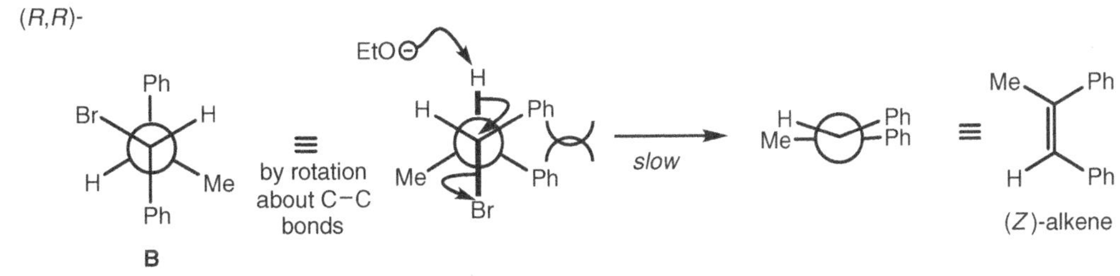

E2 elimination

$B^{\ominus}$ = EtO$^{\ominus}$; LG = Br$^{\ominus}$

(S,R)-

(R,R)-

The (E/Z)-stereoselectivity is a consequence of the antiperiplanar arrangement of the H and Br in the E2 transition state. The rate difference reflects the greater steric interactions between the two Ph groups in the transition state from (R,R)- compared with Ph and Me in the complementary (S,R)- reaction.

3. Illustrate and explain the competition between substitution and elimination by considering the reaction of NaOH with *syn-* and *anti-tert*-butylcyclohexyl tosylate **A**.

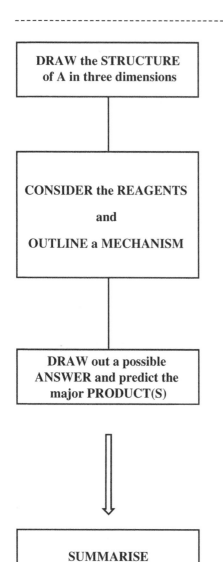

A   *t*-Bu —⟨ ⟩〜OSO$_2$—⟨ ⟩—

(where 〜 indicates both stereoisomers at this position)

DRAW the STRUCTURE of A in three dimensions	The *tert*-butyl group is an **equatorial anchor**. *The conformer with the tert-butyl **axial** has too much steric interaction with the neighbouring axial H atoms to be populated.* There will be two major conformers, one with the tosylate axial (*syn*-**A**) and the other equatorial (*anti*-**A**). *The anti-diastereoisomer will be more stable than the syn-.*
CONSIDER the REAGENTS and OUTLINE a MECHANISM	⁻OH is both a base and a nucleophile. ⁻OTs is a good leaving group. Basic conditions disfavour carbocations, so E2/S$_N$2 reactions look more likely. S$_N$1, E1 and E1$_{cb}$ mechanisms are also discounted because there are no suitable stabilising groups for either a cationic or an anionic intermediate. *E2 reactions normally have antiperiplanar transition states; for alignment of H and LG on a cyclohexane ring, both substituents must be diaxial. S$_N$2 reactions have in-line transition states and proceed with inversion of configuration.*
DRAW out a possible ANSWER and predict the major PRODUCT(S)	The choice here between E2 and S$_N$2 will depend primarily on the availability of the antiperiplanar configurations required for E2. Antiperiplanar E2 elimination from the axial *syn*-tosylate gives the alkene as the major product, with the S$_N$2 mechanism providing the minor product. E2 is difficult for the equatorial *anti*-tosylate, so S$_N$2 becomes the major pathway. *Inversion means that the equatorial tosylate gives the axial alcohol and vice versa.*
SUMMARISE	E2 elimination of the diaxial H, OTs from the *syn*-tosylate gives the alkene as the major product, whereas competitive S$_N$2 displacement gives the equatorial alcohol as the minor product. For the *anti*-tosylate, S$_N$2 gives the *syn*-alcohol as the major product because the antiperiplanar stereochemistry required for E2 is difficult to achieve.

*syn-***A**

S_N2 or E2?

*anti-***A**

S_N2 or E2?

$$TsO^- + {}^+BH + RCH=CH_2 \xleftarrow[E2]{B:} RCH_2CH_2OTs \xrightarrow[S_N2]{Nu^-} RCH_2CH_2Nu + TsO^-$$

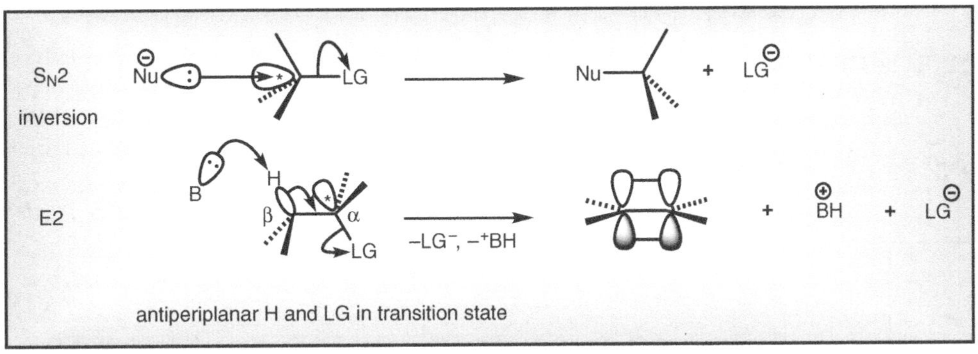

S_N2
inversion

E2

antiperiplanar H and LG in transition state

E2 needs antiperiplanar, diaxial H and OTs: only possible for *syn-*A.

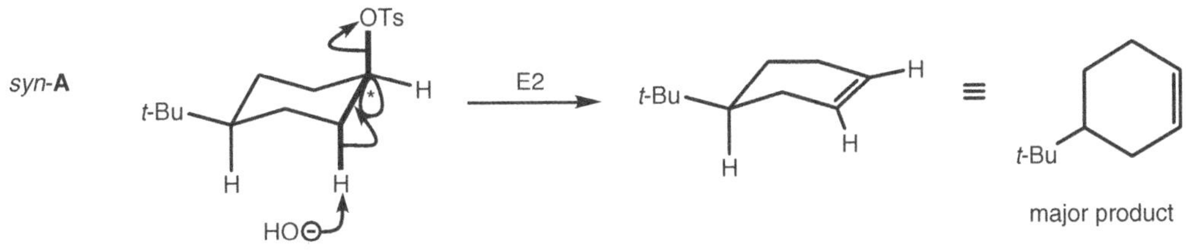

*syn-***A**

major product

*anti-***A**   OTs is equatorial, so no diaxial H and OTs:
E2 elimination is not possible via the chair conformer.

S_N2 is possible for both diastereoisomers

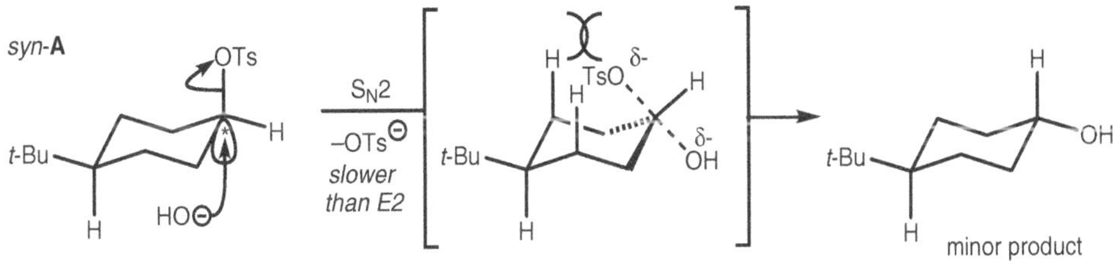

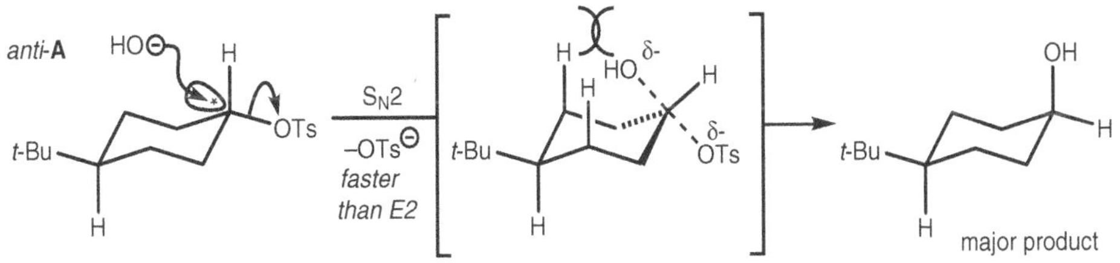

*syn-***A**

S_N2
$-OTs^{\ominus}$
*slower
than E2*

minor product

*anti-***A**

S_N2
$-OTs^{\ominus}$
*faster
than E2*

major product

4. Suggest a mechanism for the following reaction.

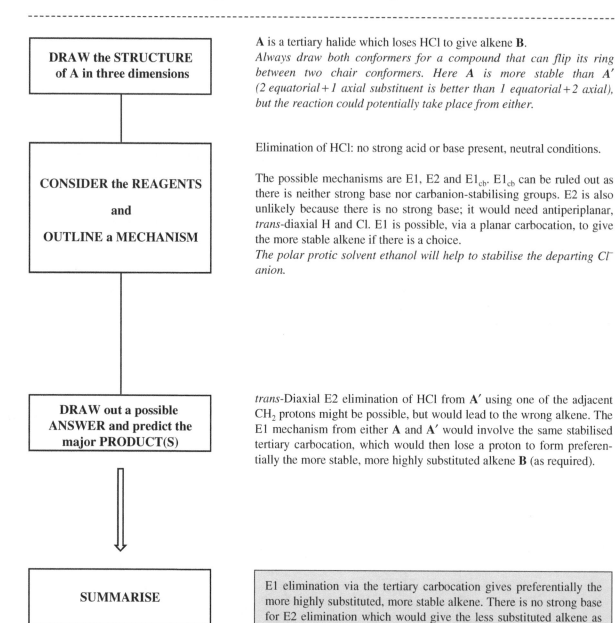

DRAW the STRUCTURE of A in three dimensions

A is a tertiary halide which loses HCl to give alkene **B**.
*Always draw both conformers for a compound that can flip its ring between two chair conformers. Here **A** is more stable than **A′** (2 equatorial + 1 axial substituent is better than 1 equatorial + 2 axial), but the reaction could potentially take place from either.*

CONSIDER the REAGENTS

and

OUTLINE a MECHANISM

Elimination of HCl: no strong acid or base present, neutral conditions.

The possible mechanisms are E1, E2 and E1$_{cb}$. E1$_{cb}$ can be ruled out as there is neither strong base nor carbanion-stabilising groups. E2 is also unlikely because there is no strong base; it would need antiperiplanar, *trans*-diaxial H and Cl. E1 is possible, via a planar carbocation, to give the more stable alkene if there is a choice.
The polar protic solvent ethanol will help to stabilise the departing Cl$^-$ anion.

DRAW out a possible ANSWER and predict the major PRODUCT(S)

trans-Diaxial E2 elimination of HCl from **A′** using one of the adjacent CH$_2$ protons might be possible, but would lead to the wrong alkene. The E1 mechanism from either **A** and **A′** would involve the same stabilised tertiary carbocation, which would then lose a proton to form preferentially the more stable, more highly substituted alkene **B** (as required).

SUMMARISE

E1 elimination via the tertiary carbocation gives preferentially the more highly substituted, more stable alkene. There is no strong base for E2 elimination which would give the less substituted alkene as product.

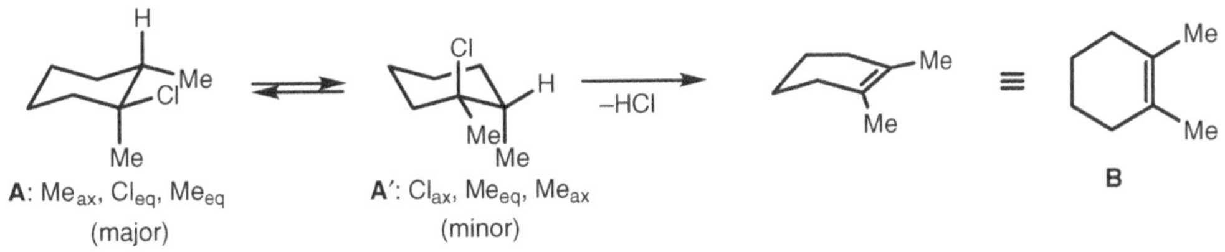
H
Me
Cl
Me
A: Me_{ax}, Cl_{eq}, Me_{eq}
(major)
Cl
H
Me
Me
A′: Cl_{ax}, Me_{eq}, Me_{ax}
(minor)
–HCl
Me
Me
≡
Me
Me
B

$$RCH_2CH(Cl)R \xrightarrow[\text{E1 or E2}]{B:} RCH{=}CHR + {}^{+}BH + Cl^{-}$$

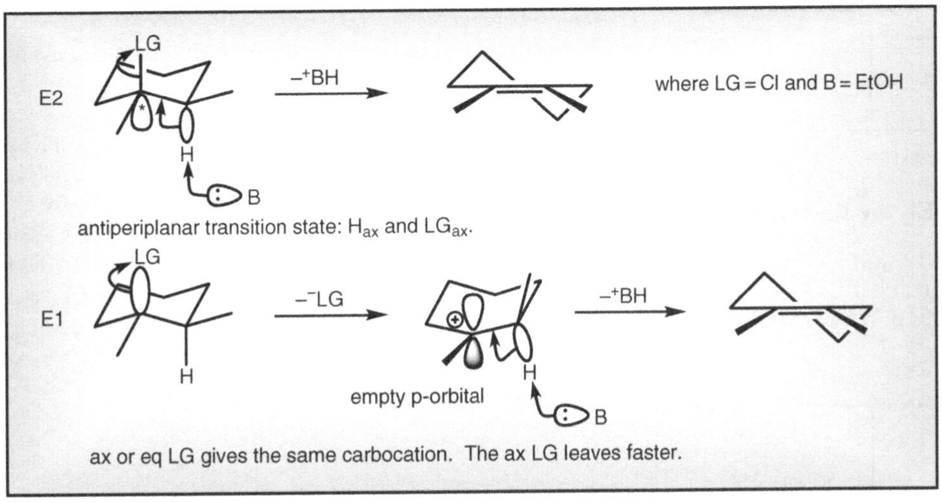
E2
LG
–⁺BH
where LG = Cl and B = EtOH
H
B
antiperiplanar transition state: H_{ax} and LG_{ax}.
E1
LG
–⁻LG
–⁺BH
H
H
empty p-orbital
B
ax or eq LG gives the same carbocation. The ax LG leaves faster.

If E2: *trans*-diaxial H, Cl

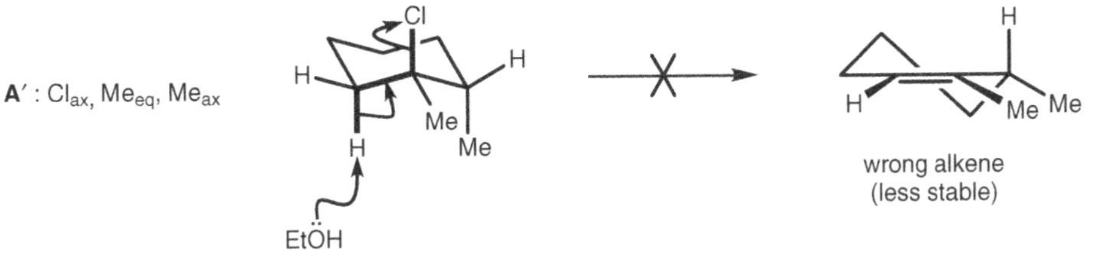
A′ : Cl_{ax}, Me_{eq}, Me_{ax}
Cl
H
H
Me
Me
H
EtÖH
H
H
Me
Me
wrong alkene
(less stable)

If E1, either **A** or **A′** gives the same carbocation.

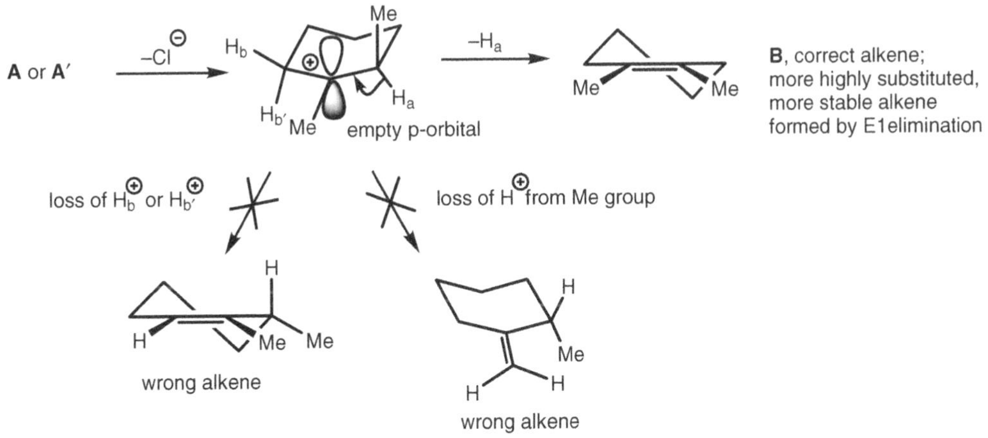
A or A′
–Cl⁻
Me
H_b
$H_{b'}$
Me
H_a
empty p-orbital
$–H_a$
Me
Me
B, correct alkene;
more highly substituted,
more stable alkene
formed by E1elimination
loss of H_b^{+} or $H_{b'}^{+}$
loss of H⁺ from Me group
H
H
Me
Me
wrong alkene
H
Me
H
H
wrong alkene

5. Suggest a suitable mechanism for the interconversion reaction of **A** to **B**.

single enantiomer
A

single enantiomer
B

[where ∿∿ indicated both diastereoisomers at this position]

DRAW the STRUCTURES of A and B in three dimensions	**A** has three chiral centres; one is unspecified so there will be two diastereo-isomers (epimers). **B** has one specified chiral centre, with stereochemistry retained from **A**. In the reaction from **A** to **B**, an epoxide becomes an allylic alcohol; the C_3–H bond is broken, a new C_3=C_2 bond is made and the C_2–O bond is broken. *Recognising a sequence of **bonds broken and bonds made** like these is very useful; it often indicates a concerted reaction.*
CONSIDER the REAGENTS and OUTLINE a MECHANISM	KOH is a base. The hydrogen atom adjacent to the aldehyde is acidic, and is easily lost. The resulting carbanion at C_3 is stabilised by delocalisation with the aldehyde (CHO) group to form an enolate. A stable carbanion can be formed next to the CHO, which could act as a nucleophile to ring open the epoxide (if HOMO–LUMO overlap is possible). Formation of the alkene would follow an $E1_{cb}$ mechanism. *Alkoxides are not usually good leaving groups, but opening an epoxide relieves the strain within the three-membered ring.*
DRAW out a possible ANSWER and predict the major PRODUCT(S)	Loss of the proton adjacent to CHO in both diastereoisomers of **A** will give the same stable, delocalised enolate **C**. This π-system can then overlap with the adjacent σ^* of the epoxide to induce opening of the ring. Protonation of the resulting alkoxide (by the solvent, MeOH) gives **B**. Since the bonds to C_1 remain untouched, the stereochemistry at C_1 is retained from **A** to **B**.
SUMMARISE	Initial deprotonation adjacent to the aldehyde group in **A** gives a resonance stabilised carbanion (enolate) which induces ring-opening of the epoxide to form the alkene **B** by an $E1_{cb}$ mechanism.

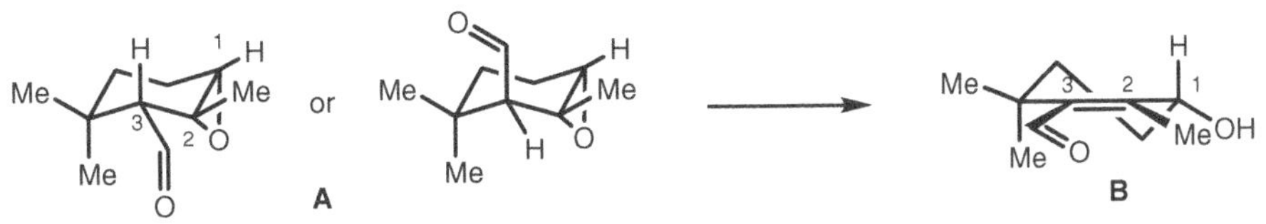
H
1
H
Me
Me
3
2
O
Me
O
or
O
H
Me
Me
Me
H
O
A
H
Me
3
2
1
Me
O
Me
OH
B

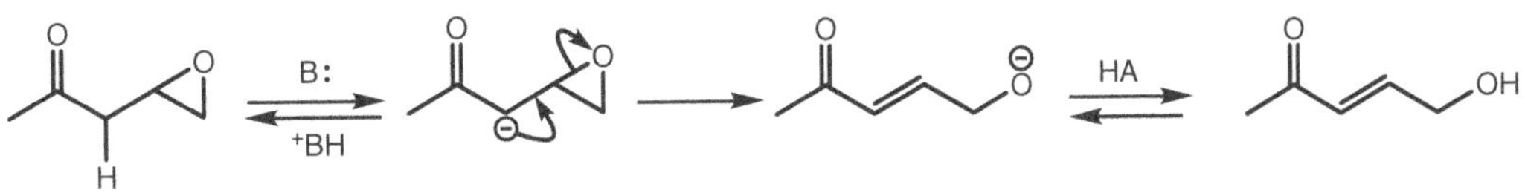
O
O
H
B:
+BH
O
O
O
O
HA
O
OH

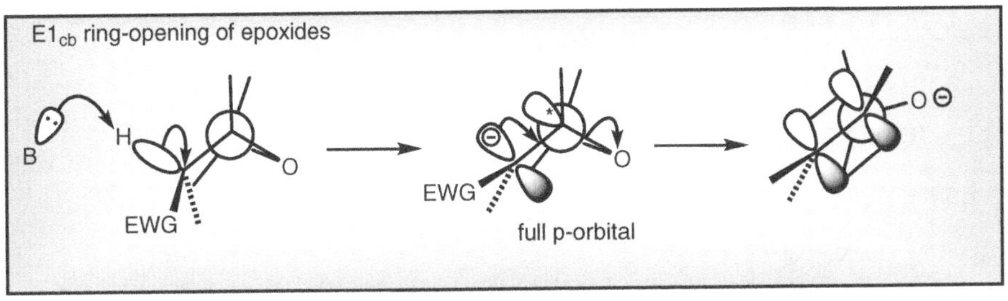
E1cb ring-opening of epoxides
B
H
O
EWG
EWG
O
full p-orbital
O

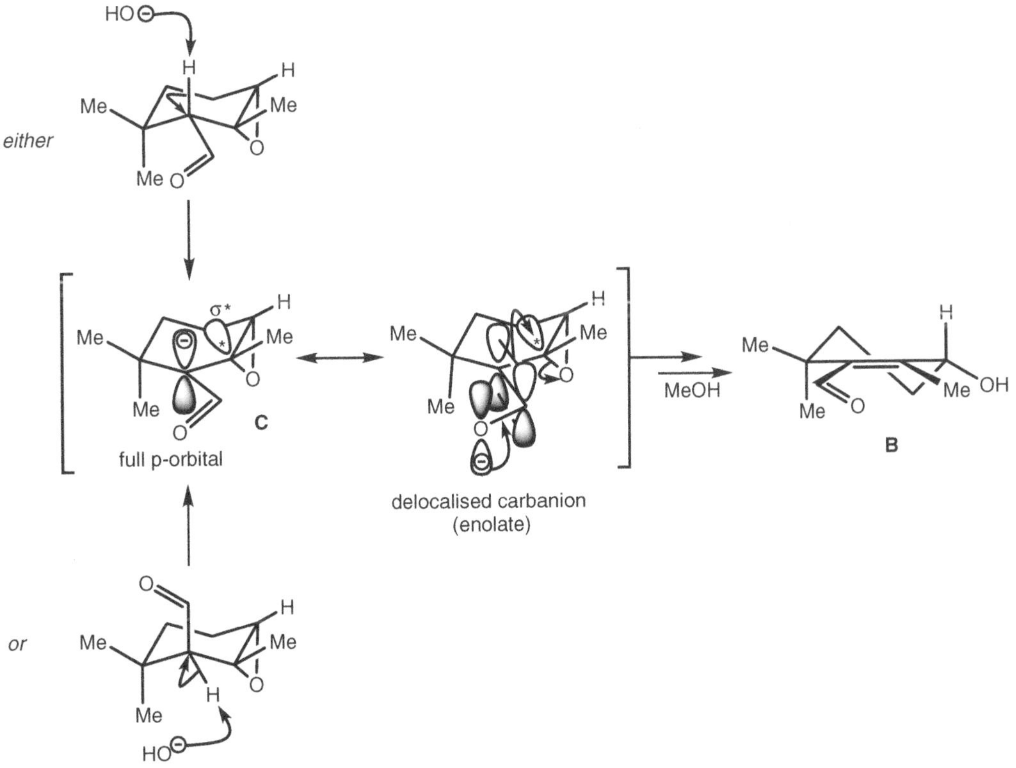
HO
H
H
Me
either
Me
Me
O
O
σ*
H
Me
Me
Me
O
O
full p-orbital
C
H
Me
Me
Me
O
O
delocalised carbanion
(enolate)
MeOH
H
Me
Me
O
Me
OH
B
or
O
H
Me
Me
Me
O
H
HO

6. Deduce the structure of the major product **B** obtained by heating sulfoxide **A**.

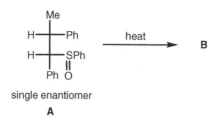

single enantiomer

A

<table>
<tr>
<td>

DRAW the STRUCTURE of A in three dimensions

</td>
<td>

A is a sulfoxide and has three chiral centres: the sulfur and two carbon atoms [top one (*R*)-, bottom one (*S*)-].
It is better to convert the given Fischer projection into either a perspective drawing or a Newman projection.
The group S=O can also be drawn as S→O and S⁺-O⁻.

</td>
</tr>
<tr>
<td>

CONSIDER the REAGENTS

and

OUTLINE a MECHANISM

</td>
<td>

No reagents: this is a pyrolysis reaction.

Pyrolysis suggests a *syn*-elimination via a cyclic transition state. The groups to be eliminated must have a synperiplanar relationship. When heated, sulfoxides eliminate sulphenic acids, RSOH, to give alkenes; the chirality at sulfur can be ignored.

</td>
</tr>
<tr>
<td>

DRAW out a possible ANSWER and predict the major PRODUCT(S)

</td>
<td>

The SOPh and H need to be synperiplanar for cyclic elimination to occur. Conformer **A** does not have this alignment, but **C** does. Cyclic elimination of PhSOH stereospecifically leads to the (*Z*)-alkene.

</td>
</tr>
<tr>
<td>

SUMMARISE

</td>
<td>

Thermal *syn*-elimination of **A** occurs stereospecifically via an internal synperiplanar cyclic process giving the (*Z*)-alkene **B**.

</td>
</tr>
</table>

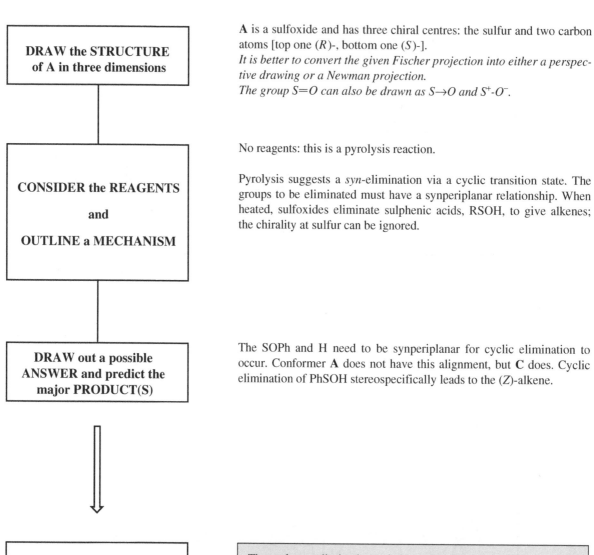

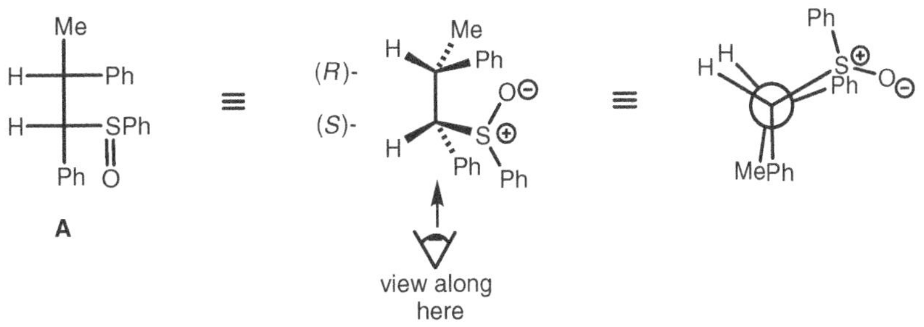
Me
H
Ph
H
SPh
Ph
O
A
(R)-
(S)-
view along here

$$PhSOCH_2CH_2R \xrightarrow{\text{heat}} CH_2{=}CHR + PhSOH$$

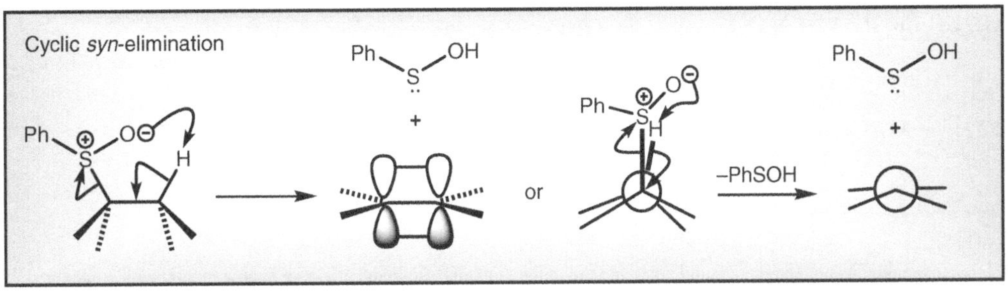
Cyclic syn-elimination
Ph
OH
or
–PhSOH

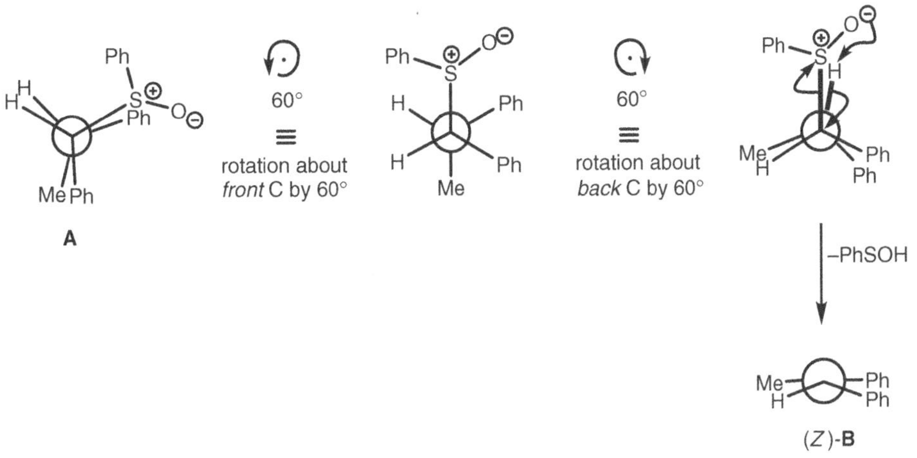
A
60°
rotation about front C by 60°
60°
rotation about back C by 60°
–PhSOH
(Z)-B

Or if you prefer perspective drawings:

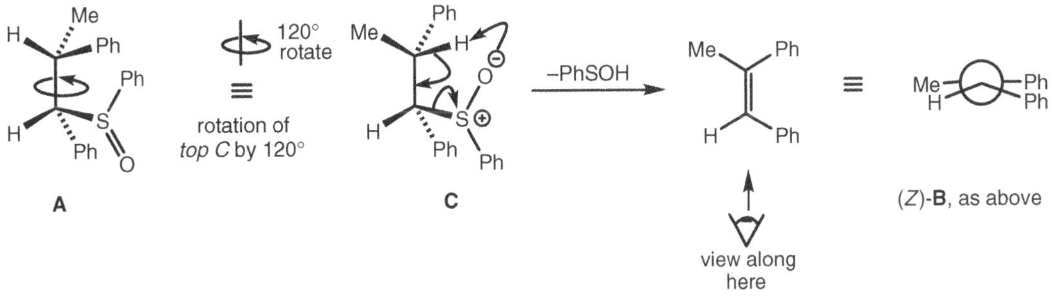
A
120° rotate
rotation of top C by 120°
C
–PhSOH
view along here
(Z)-B, as above

7. Suggest structures for **B** and **C** below, giving your reasoning.

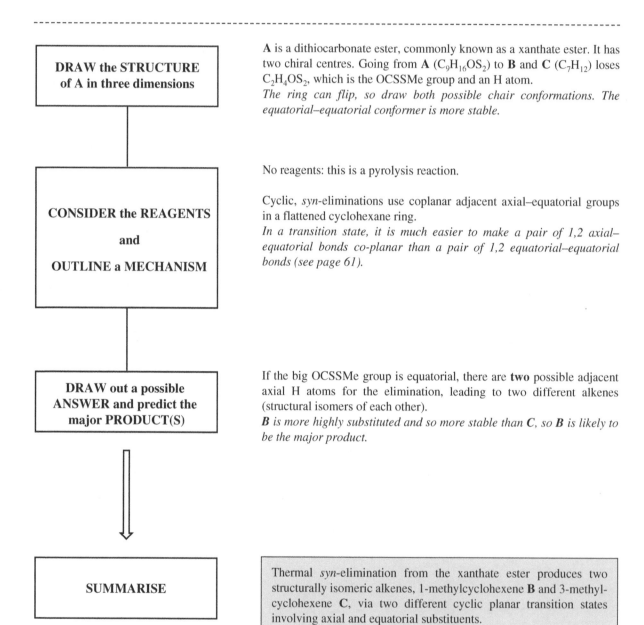

<table>
<tr><td>

DRAW the STRUCTURE of A in three dimensions

</td><td>

A is a dithiocarbonate ester, commonly known as a xanthate ester. It has two chiral centres. Going from **A** ($C_9H_{16}OS_2$) to **B** and **C** (C_7H_{12}) loses $C_2H_4OS_2$, which is the OCSSMe group and an H atom.
The ring can flip, so draw both possible chair conformations. The equatorial–equatorial conformer is more stable.

</td></tr>
<tr><td>

CONSIDER the REAGENTS

and

OUTLINE a MECHANISM

</td><td>

No reagents: this is a pyrolysis reaction.

Cyclic, *syn*-eliminations use coplanar adjacent axial–equatorial groups in a flattened cyclohexane ring.
In a transition state, it is much easier to make a pair of 1,2 axial–equatorial bonds co-planar than a pair of 1,2 equatorial–equatorial bonds (see page 61).

</td></tr>
<tr><td>

DRAW out a possible ANSWER and predict the major PRODUCT(S)

</td><td>

If the big OCSSMe group is equatorial, there are **two** possible adjacent axial H atoms for the elimination, leading to two different alkenes (structural isomers of each other).
B is more highly substituted and so more stable than C, so B is likely to be the major product.

</td></tr>
<tr><td>

SUMMARISE

</td><td>

Thermal *syn*-elimination from the xanthate ester produces two structurally isomeric alkenes, 1-methylcyclohexene **B** and 3-methylcyclohexene **C**, via two different cyclic planar transition states involving axial and equatorial substituents.

</td></tr>
</table>

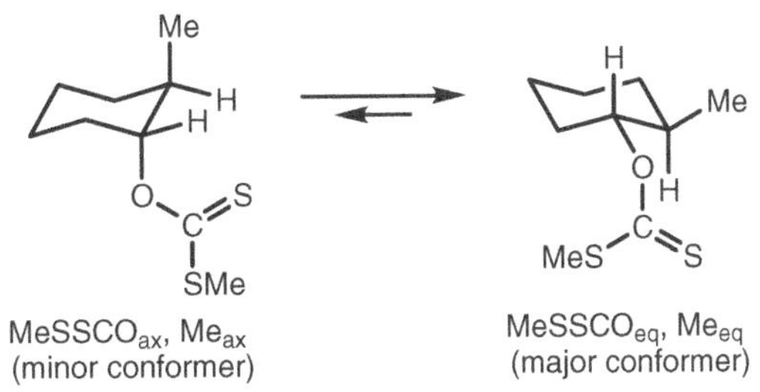

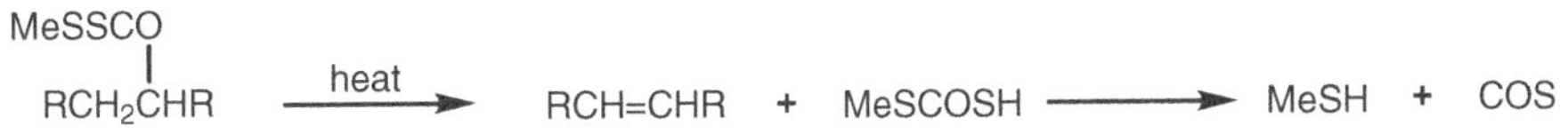

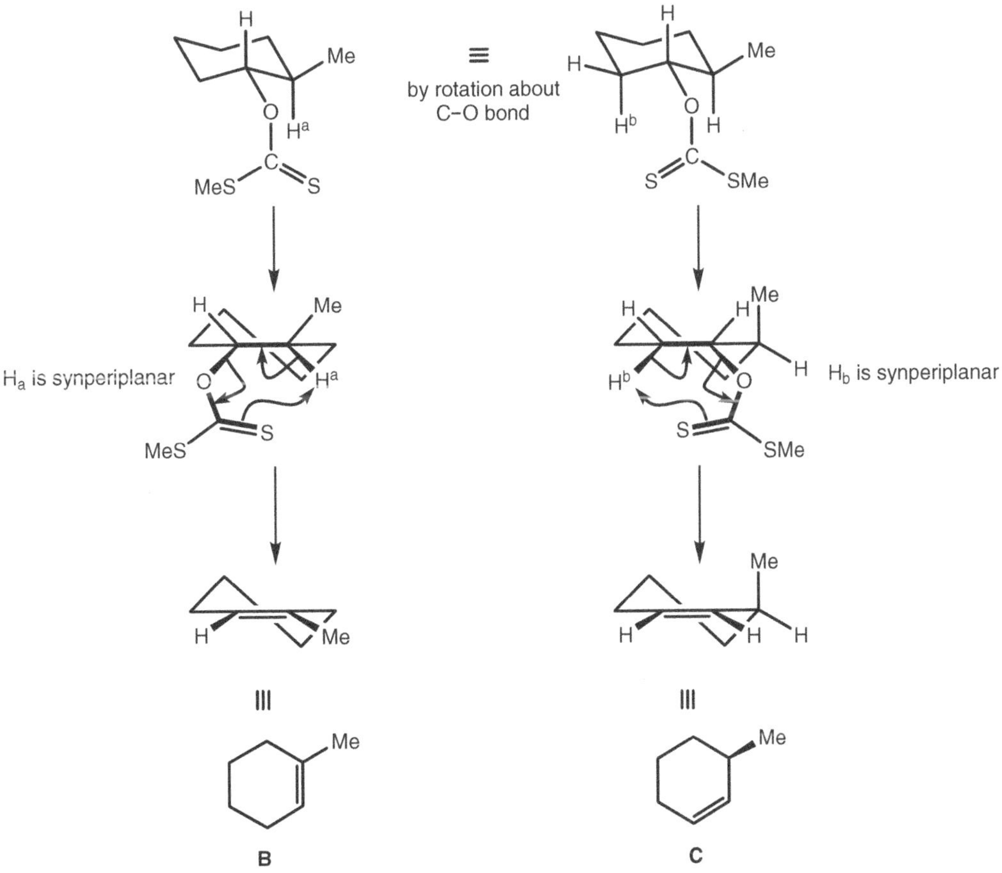

Me
H
H
O
C
S
SMe
H
Me
O
H
C
MeS
S
MeSSCOax, Meax
(minor conformer)
MeSSCOeq, Meeq
(major conformer)

MeSSCO
RCH2CHR
heat
RCH=CHR
+
MeSCOSH
MeSH
+
COS

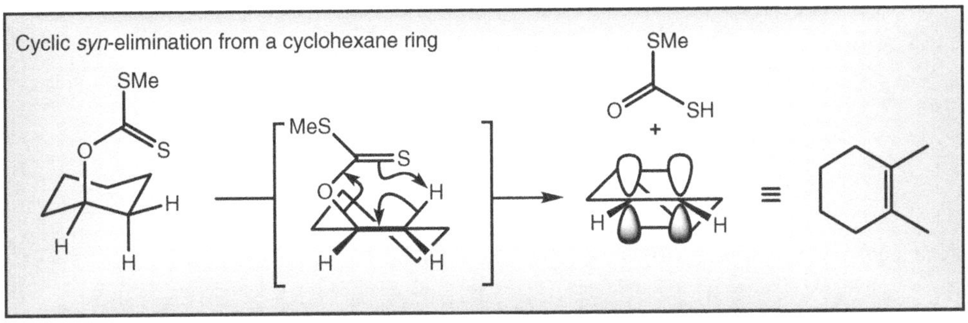

Cyclic syn-elimination from a cyclohexane ring
SMe
O
S
H
H
H
MeS
S
O
H
H
H
SMe
O
SH
+
H
H
≡

H
Me
O
Ha
C
MeS
S
≡
by rotation about
C–O bond
H
H
Me
Hb
O
H
S
C
SMe
H
Me
Ha is synperiplanar
O
Ha
MeS
S
H
H
Me
Hb
H
O
Hb is synperiplanar
S
SMe
H
Me
Me
H
H
H
Me
Me
B
C

Two isomeric alkenes **B** and **C** are formed

Additional problems for Chapter 3

1. Elimination of HCl from dichloride **A** is much faster than from dichloride **B**. Comment on this observation.

A **B**

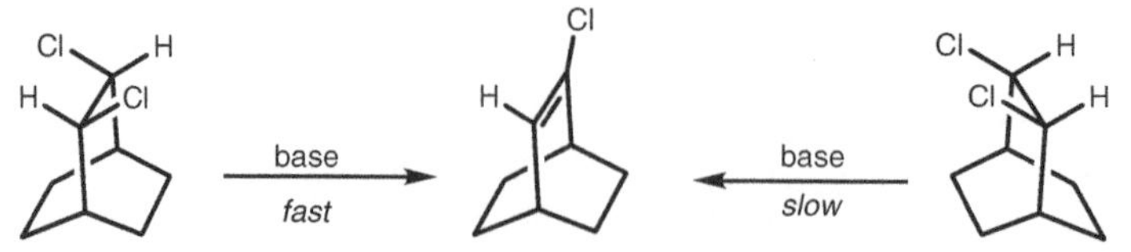

2. Rationalise the different modes of elimination for the alkenes **C** and **D**.

C

D

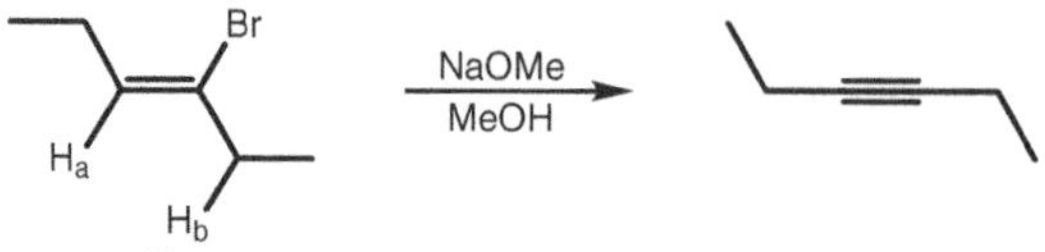

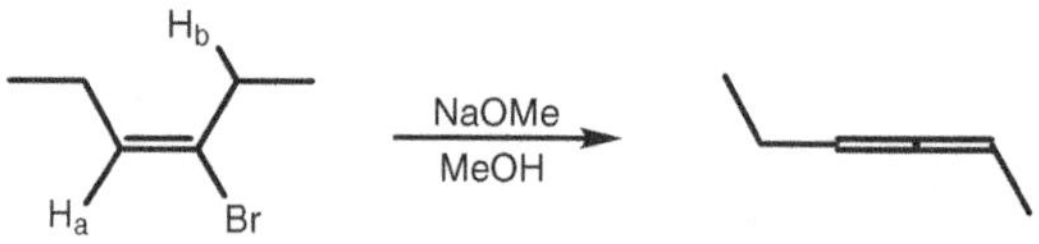

3. Explain the regio- and stereoselectivity of the following reaction.

single enantiomer **E** | **F**; 93% | single enantiomer **G**; 7%

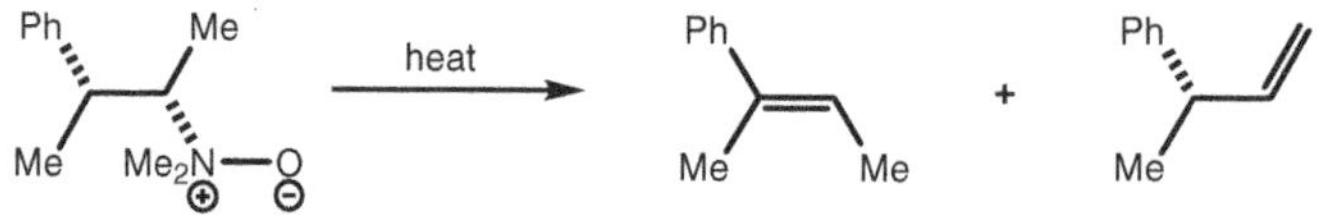

4. Explain why heating (*R,S*)-**H** gives a product containing no deuterium.

(*R,S*)-**H**

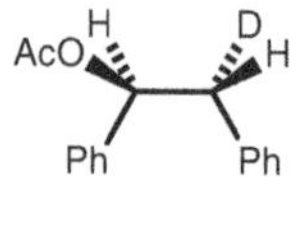

5. Suggest a mechanism for the following reaction.

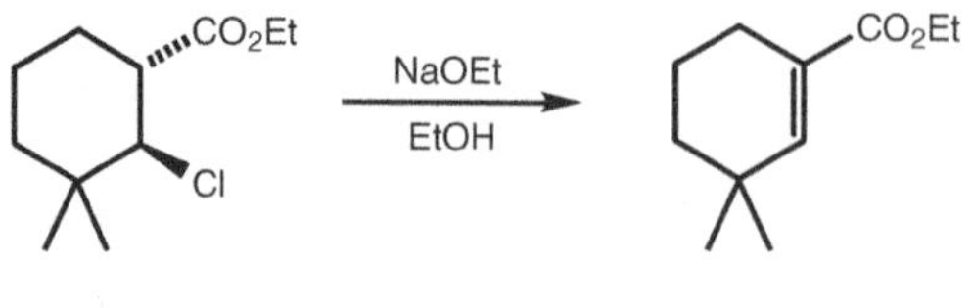

Answers: 1. E2 elimination – for **A**: synperiplanar HCl is perfectly set up (no antiperiplanar conformer available), whereas for **B**: no syn or antiperiplanar conformer is available thus disfavouring elimination; 2. E2 elimination – antiperiplanar is more favourable than synperiplanar; for **C**: H_a is preferred over H_b due to fixed conformation, whereas for **D**: antiperiplanar H_b is preferred over synperiplanar H_a; 3. Concerted cyclic synperiplanar β-elimination leads stereospecifically to **F** and **G**. Double bond in **F** is more stable (and thus preferred) due to conjugation to Ph and being triply substituted; 4. Cyclic elimination using the acetate group gives the more stable (*E*)-alkene, stilbene, and CH_3CO_2D. Loss of CH_3CO_2H would lead to the less stable and less preferred (*Z*)-alkene. Stereospecific loss of CH_3CO_2D is favoured despite breaking the stronger CD bond; 5. Elimination of HCl under basic conditions. Mechanism cannot be E2 due to the absence of an antiperiplanar H. E1 disfavoured due to poor carbocation stability. Reaction proceeds via an $E1_{cb}$ elimination pathway – by deprotonation adjacent to the C=O group followed by loss of chloride ion to give the required enone. This process can occur irrespective of the stereochemical relationship between H and Cl.

4 Electrophilic Addition to Unsaturated Compounds

> The aim of this chapter is to develop an understanding of the stereochemical features of alkenes and their reactions with electrophiles.
>
> At the end of this chapter, you should be able to:
>
> - Draw and describe the configuration of a given alkene.
> - Draw and compare the mechanisms of electrophilic addition to an alkene for four common types of electrophilic reagents: halogens, OsO_4, per-acids and boranes.
> - Explain the stereoselectivity of these reactions.

Background

The rigidity of the carbon–carbon double bond causes geometric (E/Z)-isomerism in suitably substituted alkenes (for the use of E/Z descriptors, see page 16). The π-orbital of the double bond is a centre of high electron density and acts as a nucleophile or base by reacting with electrophilic or acidic reagents. However, if the C=C bond is conjugated with an electron-withdrawing group, such as C=O, this situation is reversed and the C=C can then attract nucleophiles in the same way that a C=O does. These nucleophilic (Michael) additions are better considered as an extension to C=O addition, and will not be considered here.

Electrophilic addition to alkenes

The HOMO of an alkene is its π-orbital, with its high electron density above and below the plane. The direction of attack by electrophiles will therefore be from directly above (or below) the molecular plane. These addition reactions can occur by two related processes:

(1) Stepwise electrophilic addition – by formation of an intermediate, e.g. a carbocation or bromonium ion. This is usually followed by nucleophilic addition to give saturated compounds (e.g. addition of HBr).
(2) Concerted electrophilic addition – by formation of *two* new bonds to the electrophile to give saturated compounds (e.g. addition of OsO_4, per-acids and boranes).

Stepwise electrophilic addition to a π-bond

These reactions can proceed in two ways:

(1) By addition of an electrophilic centre which contains non-bonded electrons (e.g. Br_2). Formation of the cyclic bromonium ion is followed by *anti*-addition of Br⁻.

bromonium ion   *anti*-addition   1,2-dibromide

(2) By addition of an electrophilic centre which does not contain non-bonded electrons; this typically involves the addition of an H⁺ ion. This can result in both *syn*- and *anti*-addition of the counter-ion. For example, HBr addition to an alkene can result in the formation of two related *syn*- and *anti*-bromides.

empty p-orbital   *syn*-addition   *anti*-addition

Concerted electrophilic addition to a π-bond

These reactions characteristically proceed via stereospecific *syn*-addition. Both new bonds are attached to the same side of the molecule in the initial product; e.g. addition of osmium tetraoxide to an alkene gives an osmate ester – this is usually hydrolysed to give the *syn*-diol product.

osmate ester

1. Addition of bromine to a solution of (Z)-but-2-ene gives the racemic dibromide **A**. Account for this observation, and predict the structure of the product formed by addition of deuterium (D$_2$) to (Z)-but-2-ene in the presence of palladium-on-charcoal catalyst (Pd-C).

DRAW the STRUCTURE of (Z)-but-2-ene and dibromide A in three dimensions	But-2-ene has two geometric isomers; the (Z)-isomer has the high priority Me groups on the same side of the alkene. 1,2-Dibromobutane has three stereoisomers: two enantiomers and an achiral *meso*-compound.

CONSIDER the REAGENTS and OUTLINE a MECHANISM

The but-2-ene C=C double bond is electron rich. Both bromine and deuterium will add to electron-rich alkenes.
Deuterium is an isotope of hydrogen and it will react in the same way as hydrogen.

Bromine will add via the bromonium ion mechanism; there is not enough stabilisation to favour the alternative carbocation intermediate, BrCH(Me)HMeC$^+$. S$_N$2 substitution by the bromide ion on the bromonium ion gives the 1,2-dibromide with overall *anti*-addition.
A polar mechanism is expected because there are no indications of a radical process (e.g. irradiation or radical initiators). Close comparison of the structures of the alkene and the dibromide (re-drawn if necessary) shows you that the bromines came from opposite faces. The palladium catalysed deuterium addition is a syn-process. No detailed mechanism is expected – but a schematic representation is outlined to account for the syn-selectivity.

DRAW out a possible ANSWER and predict the major PRODUCT(S)

This alkene has a plane of symmetry; attack by Br$_2$ on the top (or bottom) face gives the same bromonium ion. The bromonium ion also has a plane of symmetry. S$_N$2 at either carbon is equally likely and the racemic dibromide **A** is formed. Overall, this results in *anti*-addition.
Note that the stereoisomeric meso-dibromide is not formed.
syn-Addition of D$_2$ across the C=C bond gives the symmetrical *meso*-stereoisomer of 2,3-dideuteriobutane.
This time it is the racemic stereoisomer that is not formed; these two reactions show opposite stereoselectivity, one anti- and the other syn-.

SUMMARISE

Br$_2$ and D$_2$ addition to an alkene occur via two distinct mechanisms. Stereochemically, Br$_2$ adds *anti-* (giving the racemic dibromide), whereas D$_2$ prefers *syn*-addition. Palladium catalysed addition of D$_2$ to (Z)-but-2-ene gives the *meso-* form of 2,3-dideuteriobutane.

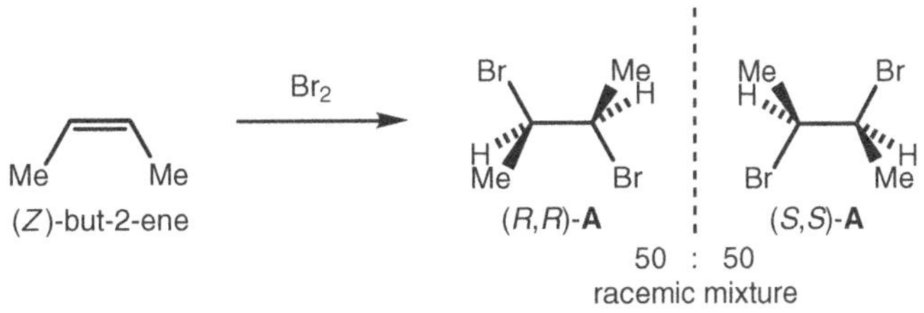

(Z)-but-2-ene $\xrightarrow{Br_2}$ (R,R)-**A** (S,S)-**A**

50 : 50
racemic mixture

$CH_2{=}CH_2 + Br_2 \longrightarrow BrCH_2CH_2Br$

$CH_2{=}CH_2 + D_2 \xrightarrow{Pd\text{-on-C}} DCH_2CH_2D$

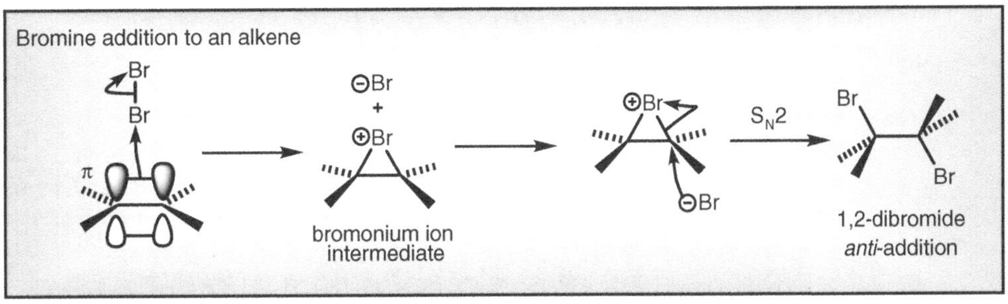

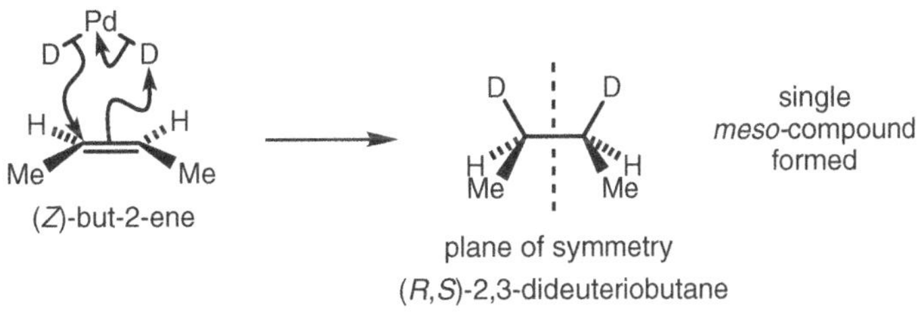

Bromine addition to an alkene

bromonium ion
intermediate

1,2-dibromide
anti-addition

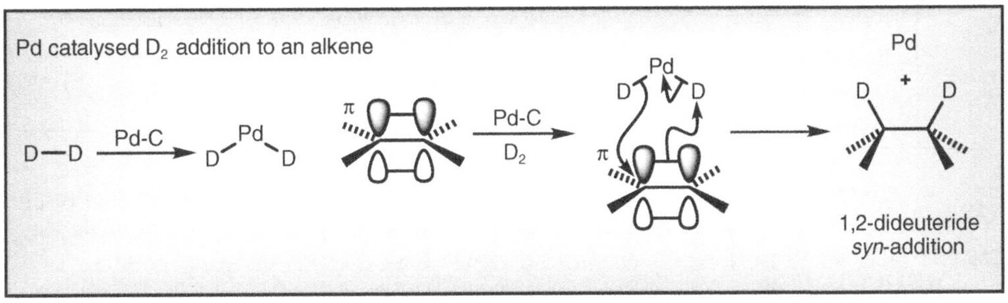

Pd catalysed D₂ addition to an alkene

1,2-dideuteride
syn-addition

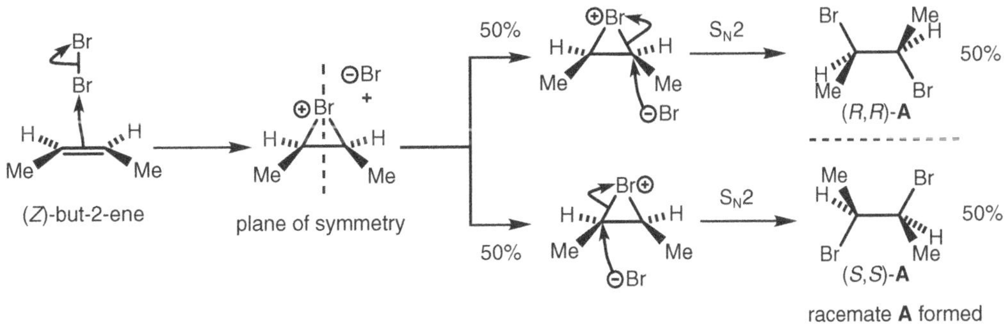

(Z)-but-2-ene   plane of symmetry   (R,R)-**A**   50%

(S,S)-**A**   50%

racemate **A** formed

(Z)-but-2-ene

single
meso-compound
formed

plane of symmetry
(R,S)-2,3-dideuteriobutane

2. Dihydroxylation of a mixture of isomeric but-2-enes with OsO_4 gives a mixture of two diastereoisomeric 1,2-diols, **A** and **B**. Draw the structure of these products and rationalise their formation.

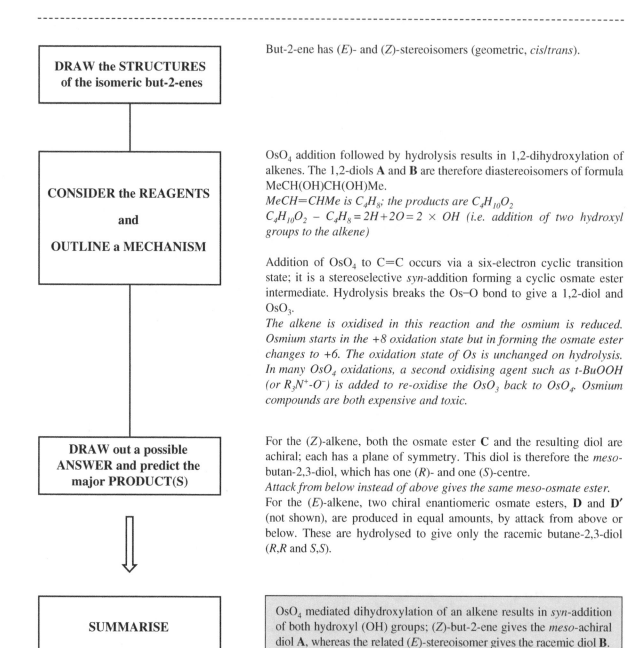

DRAW the STRUCTURES of the isomeric but-2-enes

But-2-ene has (E)- and (Z)-stereoisomers (geometric, *cis/trans*).

CONSIDER the REAGENTS and OUTLINE a MECHANISM

OsO_4 addition followed by hydrolysis results in 1,2-dihydroxylation of alkenes. The 1,2-diols **A** and **B** are therefore diastereoisomers of formula MeCH(OH)CH(OH)Me.
MeCH=CHMe is C_4H_8; the products are $C_4H_{10}O_2$
$C_4H_{10}O_2 - C_4H_8 = 2H + 2O = 2 \times OH$ *(i.e. addition of two hydroxyl groups to the alkene)*

Addition of OsO_4 to C=C occurs via a six-electron cyclic transition state; it is a stereoselective *syn*-addition forming a cyclic osmate ester intermediate. Hydrolysis breaks the Os–O bond to give a 1,2-diol and OsO_3.
The alkene is oxidised in this reaction and the osmium is reduced. Osmium starts in the +8 oxidation state but in forming the osmate ester changes to +6. The oxidation state of Os is unchanged on hydrolysis. In many OsO_4 oxidations, a second oxidising agent such as t-BuOOH (or R_3N^+-O^-) is added to re-oxidise the OsO_3 back to OsO_4. Osmium compounds are both expensive and toxic.

DRAW out a possible ANSWER and predict the major PRODUCT(S)

For the (Z)-alkene, both the osmate ester **C** and the resulting diol are achiral; each has a plane of symmetry. This diol is therefore the *meso*-butan-2,3-diol, which has one (R)- and one (S)-centre.
Attack from below instead of above gives the same meso-osmate ester.
For the (E)-alkene, two chiral enantiomeric osmate esters, **D** and **D'** (not shown), are produced in equal amounts, by attack from above or below. These are hydrolysed to give only the racemic butane-2,3-diol (R,R and S,S).

SUMMARISE

OsO_4 mediated dihydroxylation of an alkene results in *syn*-addition of both hydroxyl (OH) groups; (Z)-but-2-ene gives the *meso*-achiral diol **A**, whereas the related (E)-stereoisomer gives the racemic diol **B**.

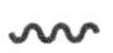

 represents undefined stereochemistry (a mixture of isomeric (*E*)- and (*Z*)-alkenes)

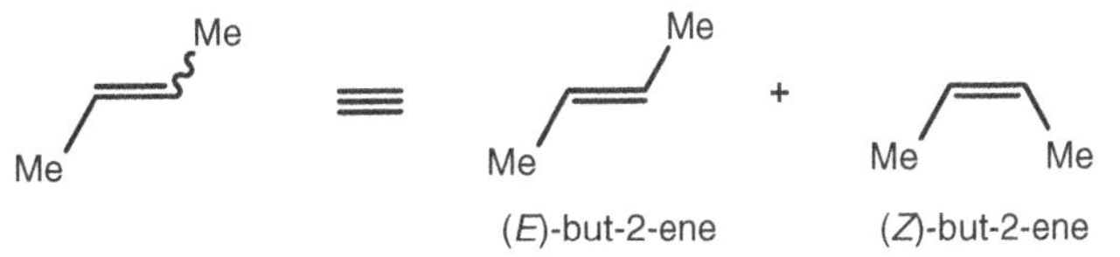
Me
Me
≡
Me
Me
+
Me
Me
(*E*)-but-2-ene
(*Z*)-but-2-ene

$$CH_2{=}CH_2 + OsO_4 + H_2O \longrightarrow HOCH_2CH_2OH + OsO_3$$

Osmium tetraoxide addition to an alkene: dihydroxylation

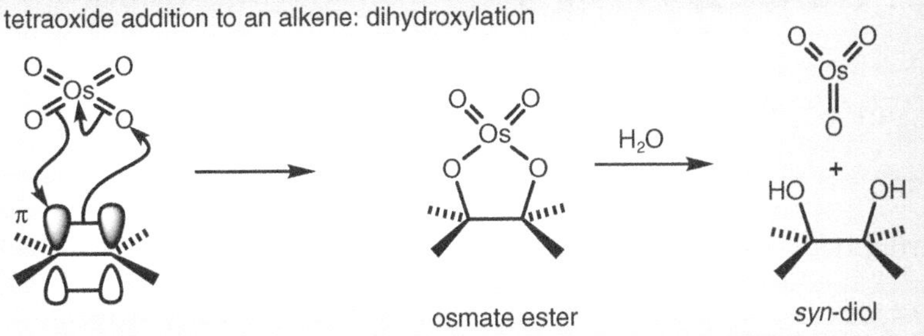
π
H_2O
+
HO
OH
osmate ester
syn-diol

Formation of the osmate ester occurs via a five-membered six-electron transition state.

For (Z) but-2-ene:

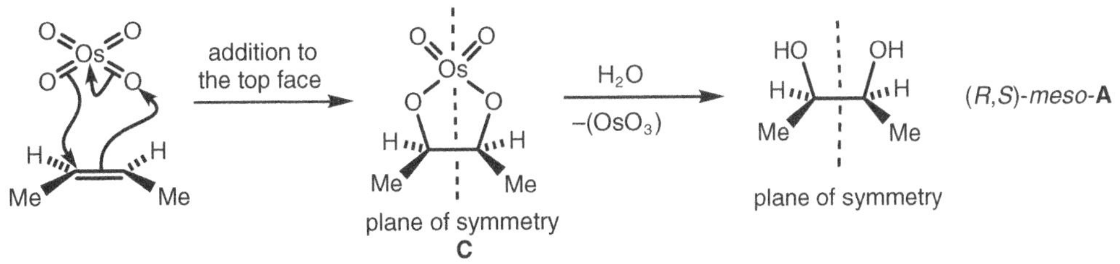
addition to
the top face
H_2O
$-(OsO_3)$
plane of symmetry
C
plane of symmetry
(*R*,*S*)-*meso*-**A**

For (*E*)-but-2-ene:

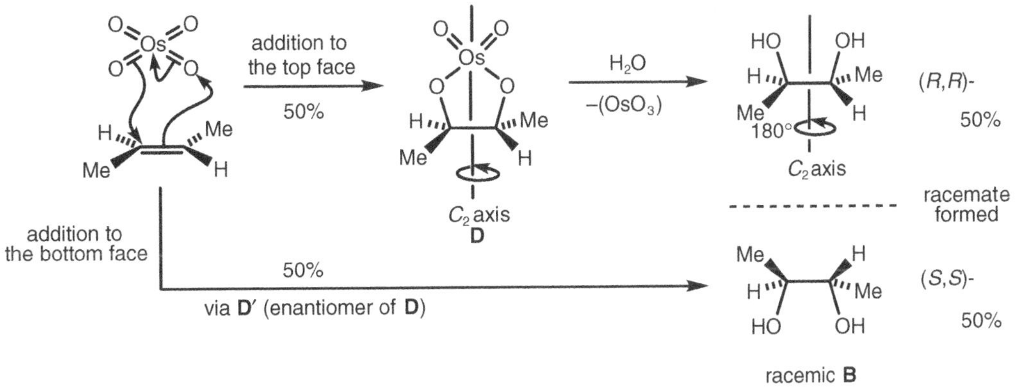
addition to
the top face
50%
H_2O
$-(OsO_3)$
(*R*,*R*)-
50%
180°
C_2 axis
C_2 axis
D
racemate
formed
addition to
the bottom face
50%
via **D′** (enantiomer of **D**)
(*S*,*S*)-
50%
racemic **B**

3. Suggest reagents for each of the following reactions and rationalise the stereochemical outcome.

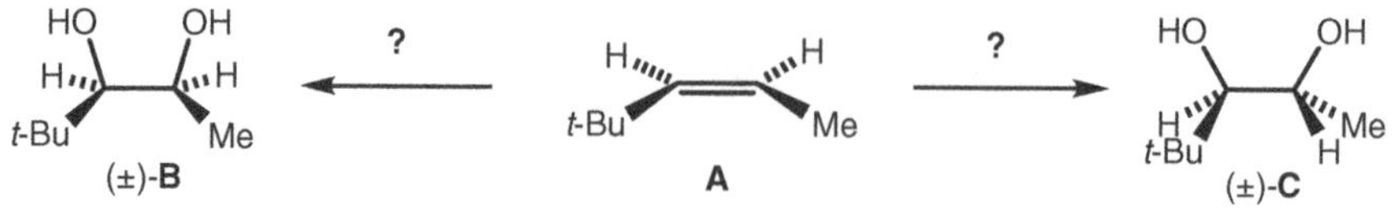

DRAW the STRUCTURE of A and relate it to B and C	Draw **A** to show its (Z)-stereochemistry, and then draw **B** and **C** to show their stereochemical relationship to **A**. *The re-drawing of A, B and C is used to indicate the stereochemistry of reactions needed. This is the key to a quick and correct answer and is very helpful in getting on the right track.*
CONSIDER the REAGENTS and OUTLINE a MECHANISM	For the conversion of **A** to **B**: a *syn*-dihydroxylation is required. Use OsO_4 then hydrolyse with water. *Perhaps it is better to use, for example, t-BuOOH and a catalytic amount of OsO₄ (see page 80).* For the conversion of **A** to **C**: *anti*-dihydroxylation is required. This is done in two stages: (i) epoxidation, using per-acid, and (ii) S_N2 ring-opening of the epoxide by hydroxide addition ($^-OH/H_2O$) to give the diol. *Stage (ii) can also be done using H⁺/H₂O.*
DRAW out a possible ANSWER and predict the major PRODUCT(S)	The osmate ester **E** is chiral; an equal amount of the other enantiomer of **E** will be made by approach of the OsO_4 from below the plane in **D**. Hydrolysis of the racemic osmate ester gives racemic diol **B**. Similarly, the epoxide **F** is chiral; both enantiomers will be made and the diol **C** will be racemic. The S_N2 ring-opening by HO⁻ takes place preferentially at the less hindered side of the two epoxide carbon atoms, at the C−Me rather than C−*t*-Bu.
SUMMARISE	Dihydroxylation of an alkene can proceed in two ways; OsO_4 mediated dihyroxylation of **A** results in *syn*-hydroxyl addition to give diol **B**, whereas the complementary *anti*-addition to **A** can be obtained by epoxidation (using a per-acid) followed by hydroxide catalysed ring-opening to give the diol **C**.

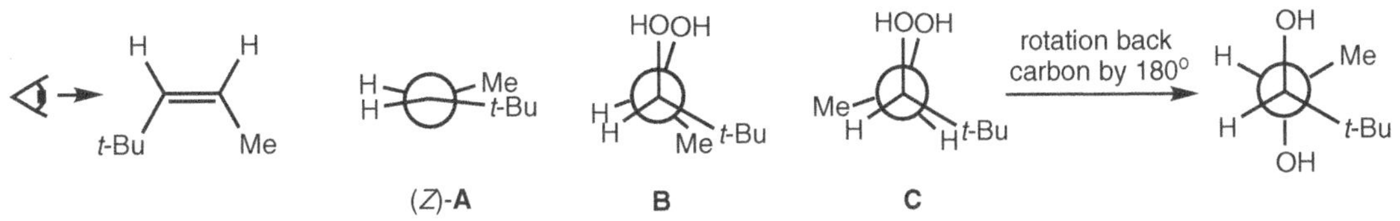

(Z)-**A**   **B**   **C**

$$CH_2=CH_2 \ + \ OsO_4 \ + \ H_2O \ \longrightarrow \ HOCH_2CH_2OH \ + \ OsO_3$$

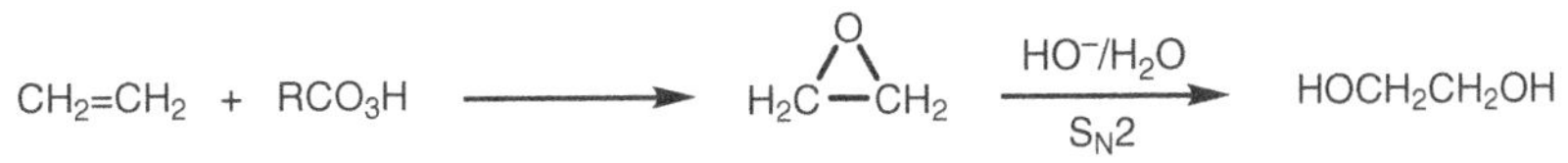

$$CH_2=CH_2 \ + \ RCO_3H \ \longrightarrow \ H_2C\overset{O}{-}CH_2 \ \xrightarrow[S_N2]{HO^-/H_2O} \ HOCH_2CH_2OH$$

Osmium tetraoxide: general mechanism – see page 81.

Epoxidation – addition of a peracid to an alkene

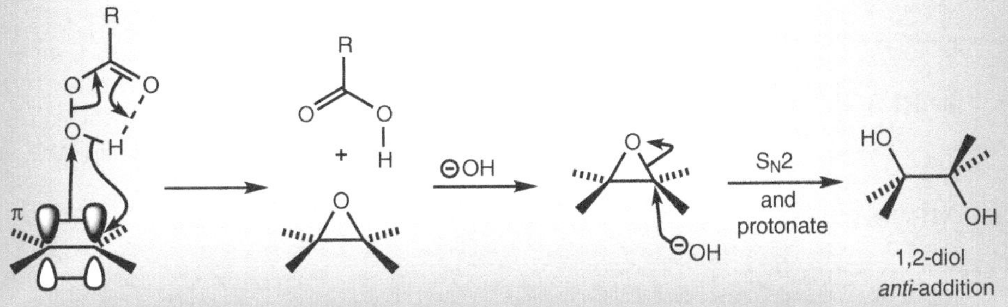

1,2-diol
anti-addition

syn-dihydroxylation – osmium tetraoxide addition

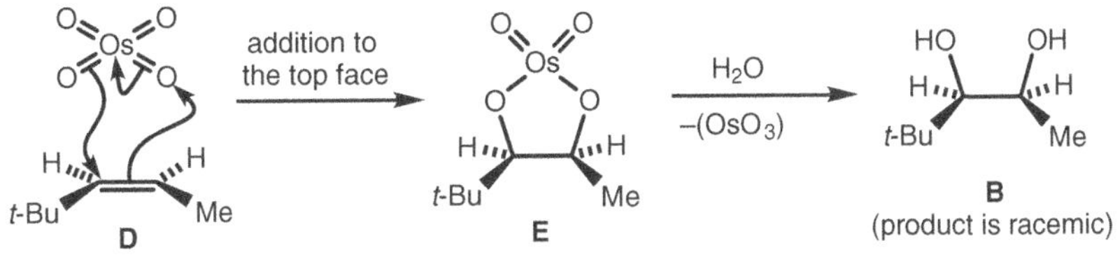

B
(product is racemic)

anti-dihydroxylation – by base-catalysed ring-opening of an epoxide with water

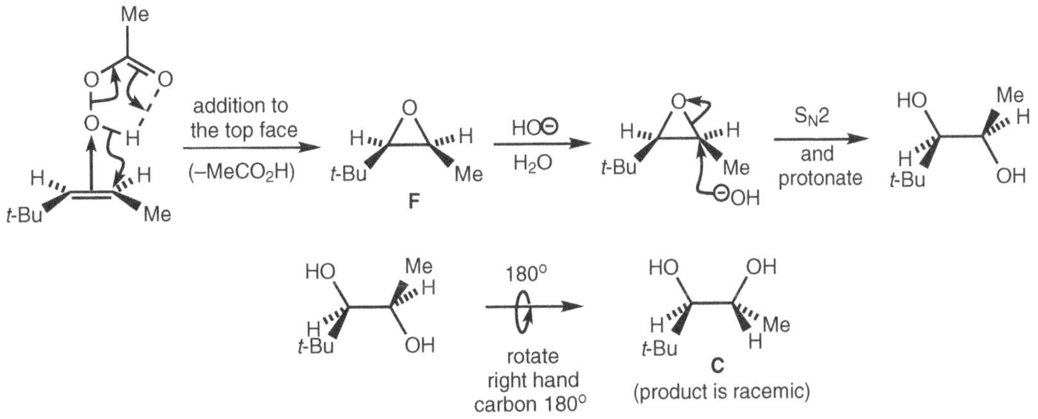

C
(product is racemic)

4. Account for the difference in stereoselectivity on epoxidation of alkenes **A** and **B** with 3-chloroperbenzoic acid (*m*-CPBA).

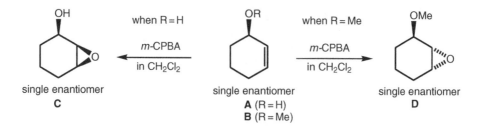

single enantiomer
C

single enantiomer
A (R = H)
B (R = Me)

single enantiomer
D

DRAW the STRUCTURES of A, B, C and D in three dimensions	The top and bottom faces of **A** and **B** are different (diastereotopic); the reaction is facially selective. Draw out **A** and **C** and then **B** and **D** to show the stereochemical relationship between each pair. *In **A** and **B**, the OH and OMe could occupy either pseudoaxial or pseudoequatorial positions next to the C=C bond. Since these are larger groups than an H atom, they will tend to occupy the less crowded pseudoequatorial positions. The stereochemistry of the epoxides is similar.*
CONSIDER the REAGENTS and OUTLINE a MECHANISM	*m*-CPBA is a per-acid which oxidises alkenes to form epoxides with *syn*-stereospecificity. This type of reaction is normally expected to occur on the less hindered face of the alkene, that is for **A** and **B** from below (on the same side as H), avoiding the bigger OH and OMe groups.
DRAW out a possible ANSWER and predict the major PRODUCT(S)	For **B** to **D**, epoxidation occurs on the less hindered face as expected. For **A** to **C**, epoxidation is on the more hindered face. This suggests that some prior coordination of reagent and substrate occurs. The OH group of **A** can act as a hydrogen bonding acceptor and donor group but the OMe of **B** can only be an acceptor. Perhaps the OH of **A** forms a hydrogen-bonded complex with the incoming per-acid which directs epoxidation on its own (upper) face of the alkene. *Prior complexation of reagent and substrate can lower the activation energy by a substantial amount – as in many enzyme-catalysed reactions. The exact pattern of the hydrogen bonding in this epoxidation is not known; one suggestion is shown.*
SUMMARISE	Addition of a per-acid to an alkene results in epoxide formation. Generally, addition occurs on the less hindered, more accessible alkene face (e.g. **B** to **D**). However, addition to the more hindered, sterically demanding face (in **A**) can be promoted by initial coordination to the incoming electrophile (by hydrogen bonding) to give epoxide **C**.

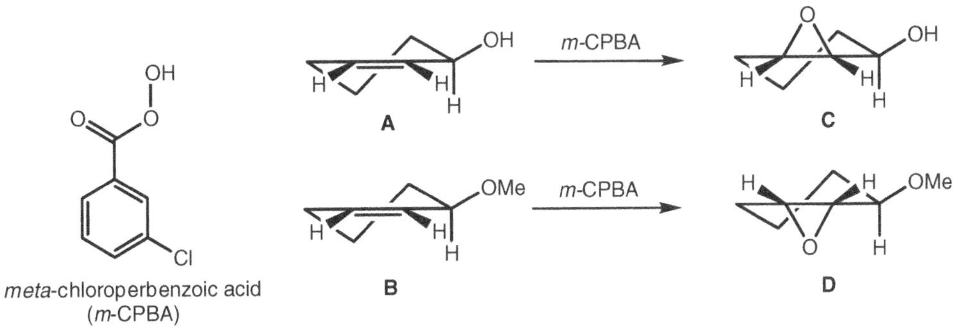

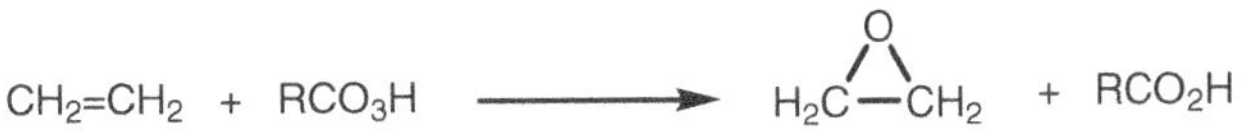

meta-chloroperbenzoic acid
(m-CPBA)

A

B

C

D

$$CH_2{=}CH_2 + RCO_3H \longrightarrow H_2C\overset{O}{\underset{}{-}}CH_2 + RCO_2H$$

Epoxidation of an alkene using a per-acid

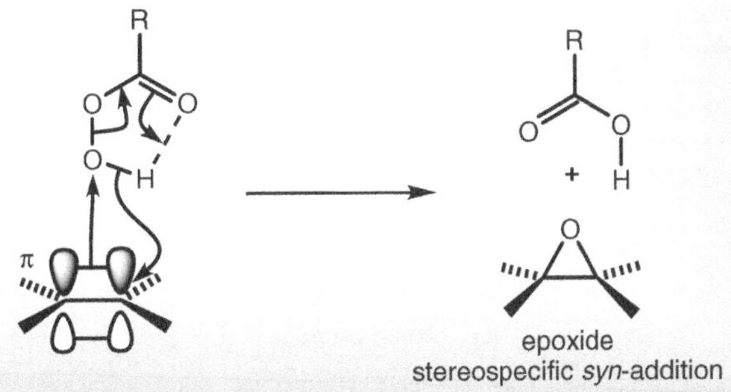

epoxide
stereospecific syn-addition

For **B** to **D**:
Epoxidation occurs on the less hindered (and less sterically demanding) face away from the methoxy group.

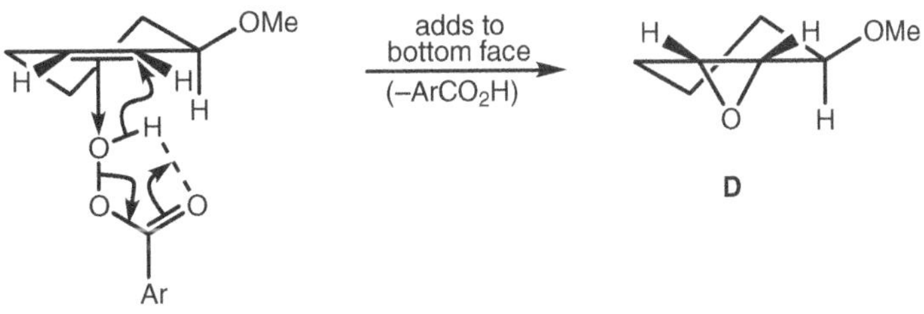

adds to
bottom face
(–ArCO₂H)

D

For **A** to **C**:
Epoxidation is directed onto the more hindered face due to hydrogen bonding from the hydroxy group.

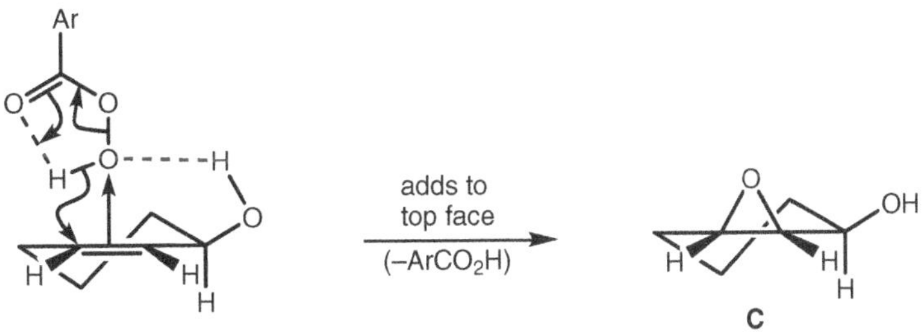

adds to
top face
(–ArCO₂H)

C

5. Hydroboration of 1-methyl cyclohexene **A** using borane (1 equivalent) gave a single diastereoisomer of 2-methyl cyclohexylborane **B**. Account for the formation of this product and rationalise the stereochemical outcome of this reaction.

1-methyl cyclohexene, **A** →[BH$_3$ in THF] 2-methyl cyclohexylborane, (±)-**B**

| DRAW the STRUCTURES of A and B in three dimensions | 2-Methyl cyclohexylborane, **B**, has two stereogenic centres and four stereoisomers: a *syn*-enantiomeric pair and an *anti*-enantiomeric pair. *Two conformers of similar energy can be drawn for syn-**B**, each with one axial and one equatorial group. For anti-**B**, the diequatorial conformer is more stable than the diaxial one.* |

| CONSIDER the REAGENTS and OUTLINE a MECHANISM | The boron atom in borane is electron deficient, sharing only six outer-shell electrons. It is electrophilic and adds to alkenes to form alkylboranes. As the new B–C bond begins to form, the boron atom bears a partial negative charge. This is shown opposite as an intermediate with full charges on boron and carbon – see inside the square brackets. If an intermediate like this does exist, it must snap shut very quickly (i.e. it has a very short lifetime) giving *syn*-addition, because the stereoselectivity of the reaction is extremely high. This reaction is both stereoselective (*syn*-) and regioselective (the boron atom adds to the side which is less crowded and the end which is less able to stabilise a positive charge). |

| DRAW out a possible ANSWER and predict the major PRODUCT(S) | Boron bonds faster to the CH end of the double bond than to the more crowded CMe end. Additionally, the CMe end stabilises a partial positive charge better. The major product *anti*-**B-1** is formed via **C** with the minor achiral product **B-2** via **D**. *The addition is syn in both cases: hydroboration has very high syn-stereoselectivity. There is no facial selectivity here because the two faces of the alkene are identical.* |

| SUMMARISE | Alkyl boranes can be formed by hydroboration of an alkene using BH$_3$. The reaction proceeds via *syn*-stereoselective addition of H and BH$_2$ across the C=C bond. BH$_2$ adds to the less hindered, more electron-rich end of the double bond. Hydroboration of the unsymmetrical alkene **A** gives the racemic *anti*-2-methylcyclohexyl borane **B**. |

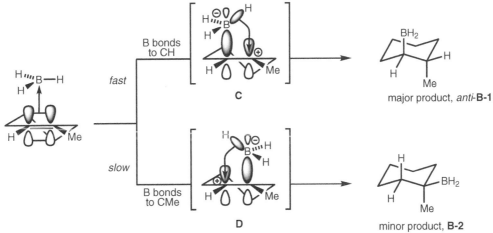

6. The scheme below gives a proposed synthesis of (Z)-1-deuteriohex-1-ene. State whether or not you think that this scheme is reasonable by working through it, showing the regio- and stereochemical consequences you expect for both stages.

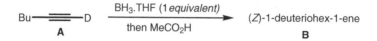

Bu————D $\xrightarrow[\text{then MeCO}_2\text{H}]{\text{BH}_3.\text{THF (1 equivalent)}}$ (Z)-1-deuteriohex-1-ene

A   B

DRAW the STRUCTURES of A and B

Draw **A** and **B** to show the similarity between them.
This indicates that overall the two H atoms have been added on the same side of the triple bond via a syn-addition process.

CONSIDER the REAGENTS and OUTLINE a MECHANISM

One equivalent of a borane adds to an alkyne to give a vinyl borane, which then reacts with a carboxylic acid replacing the boron by a hydrogen atom.
Overall two H atoms are added syn to the triple bond, one H coming from the B–H bond (in BH$_3$) and the other from the O–H bond (in MeCO$_2$H). This reduction has the same stereoselectivity as catalytic hydrogenation, but it can be used on compounds that poison catalysts, such as sulfur compounds.

DRAW out a possible ANSWER and predict the major PRODUCT(S)

BH$_3$ can add *syn* to the alkyne in two ways. The preferred regioselectivity is via the more stable, less crowded transition state to give the vinylborane (*E*)-**D** as the major product.
The less crowded transition state also has the more stable site for any partial positive charge, next to the alkyl group rather than a deuterium atom.
Protonolysis of (Z)-**C** or (E)-**D** gives the same compound, (Z)-1-deuteriohex-1-ene **B** as required.
This is because the boron is replaced by a second H which in this case is indistinguishable from the first. Since the H atoms can be added on separately (unlike catalytic hydrogenation), this process allows for syn-regiospecific addition of HD to an alkyne (see additional problems, question (5) – page 92).

SUMMARISE

Hydroboration of alkyne **A** leads to the electron-deficient vinyl borane (*E*)-**D**. This reaction proceeds via *syn*-stereoselective addition of H and BH$_2$ across the triple bond; the BH$_2$ adds to the less hindered and more nucleophilic side of the alkyne to give the vinyl borane (*E*)-**D**. Stereospecific protonolysis of this borane with acetic acid gives the required (*Z*)-1-deuteriohex-1-ene, **B**.

Bu
D

unsymmetrical alkyne

A

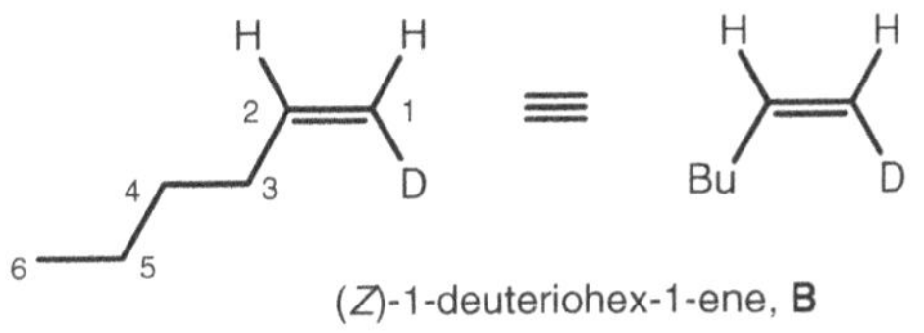
H
H
2
1
4
3
D
6
5
H
H
Bu
D

(*Z*)-1-deuteriohex-1-ene, **B**

$$RC\equiv CR \xrightarrow{BH_3} RCH{=}CRBH_2 \xrightarrow{MeCO_2H} RCH{=}CHR$$

Hydroboration of an alkyne

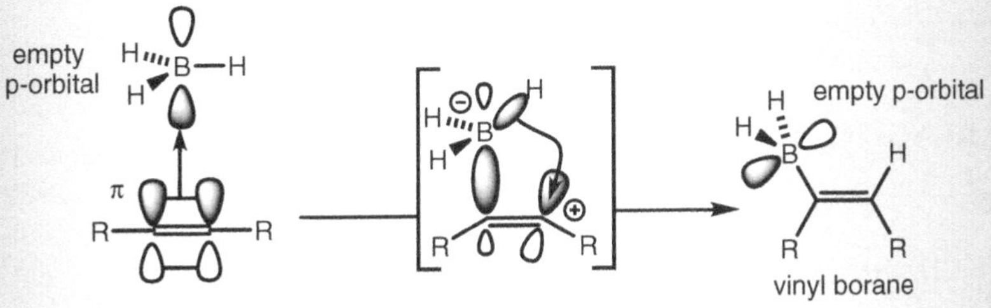
empty p-orbital
π
R
R
empty p-orbital
vinyl borane

Stereospecific *syn*-addition of borane across an alkyne

Addition:

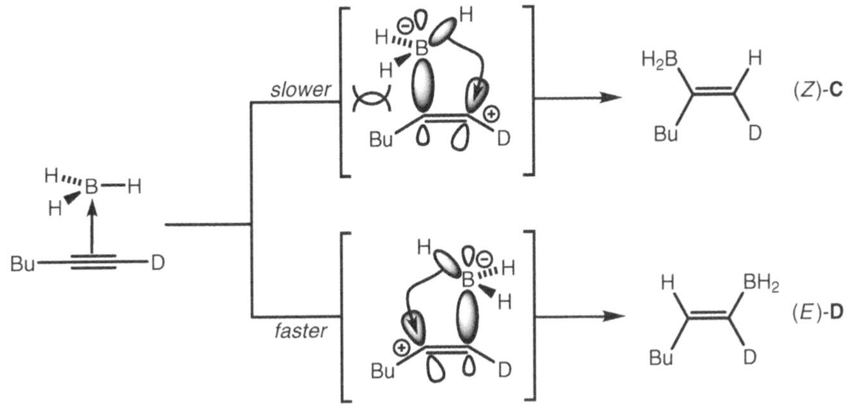
slower
faster
Bu
D
(Z)-C
(E)-D

Protonolysis

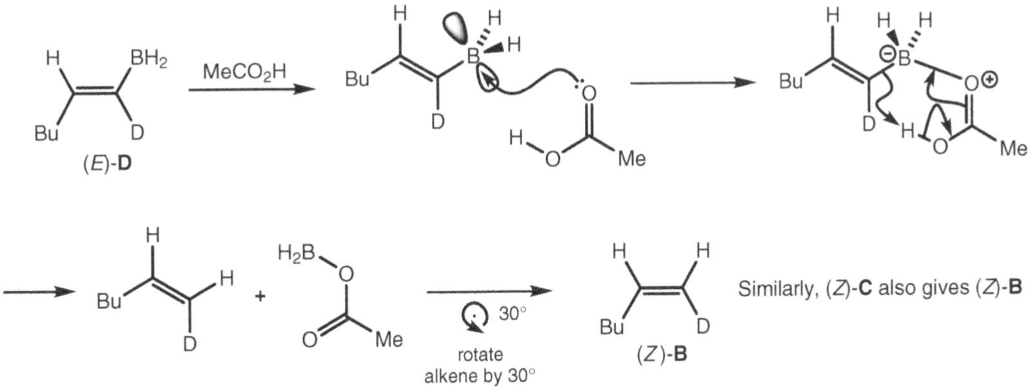
(E)-D
MeCO2H
Me
rotate alkene by 30°
30°
(Z)-B

Similarly, (*Z*)-**C** also gives (*Z*)-**B**

7. Addition of bromine water to the bicyclic alkene **A** gave the 1,2-bromoalcohol **B**. Account for the formation of **B**, and rationalise the regio- and stereochemical outcome.

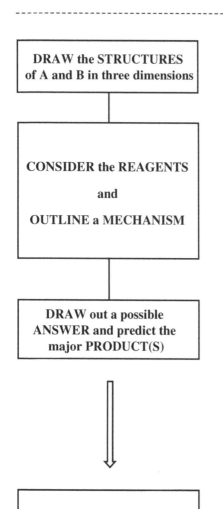

single enantiomer
A

Br₂ in H₂O →

single enantiomer
B

DRAW the STRUCTURES of A and B in three dimensions

The C=C bond of the *trans*-decalin **A** lies essentially in the plane of the two rings, with the bridge methyl above. Drawing **B** shows that the Br and OH are both axial; this suggests that they have come from different directions, top for OH and bottom for Br.

CONSIDER the REAGENTS and OUTLINE a MECHANISM

Bromine adds to an alkene to give a cyclic bromonium ion. This is then opened by water in an S_N2 reaction to give overall *anti*-addition of Br and OH.
Bromine water is acidic and contains several electrophiles, such as Br_2 and H^+. The best electrophile is probably Br_2. $BrOH_2^+$ (but not BrOH) is also good but in low concentration. Water is a reasonably good and very abundant nucleophile for the required S_N2. Final deprotonation gives the 1,2-bromoalcohol.

DRAW out a possible ANSWER and predict the major PRODUCT(S)

Approach to the top face of the alkene **A** (to give **C**) is hindered by the bridge methyl group, so bromine reacts faster on the lower face (to give **D**). The down bromonium ion **D** must undergo S_N2 opening by water from the top face, at either C_2 or C_3. Upward movement of C_2 to form a new bond to water as the C–Br bond breaks directs the reaction through a chair-like transition state, which is more favourable than that the boat-like transition state which would be formed by attack at C_3. Hence the major product is diastereoisomer **B** rather than **E**.

SUMMARISE

Addition of bromine water to the alkene **A** gives the 1,2-bromo-alcohol **B**. The reaction is stereospecific, resulting in an *anti*-addition of Br and OH. Bromination of **A** occurs on the less hindered (bottom) face (away from the axial Me group) to give the intermediate bromonium ion **D**. Ring opening with water occurs at C_2 (rather than C_3) and proceeds via a lower energy chair transition state to give **B**.

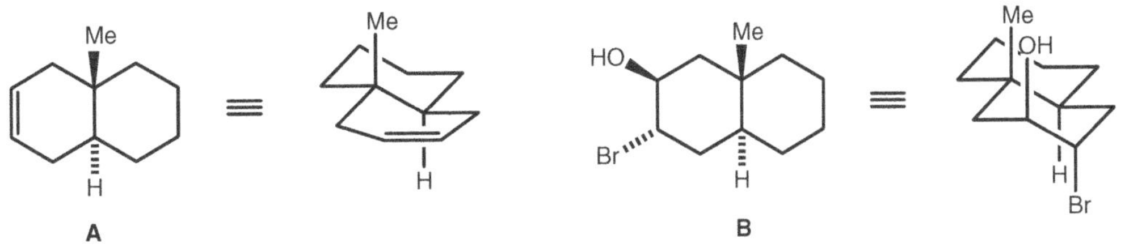
Me
H
A
HO
Me
Br
H
B
Me
OH
H
Br

$$CH_2{=}CH_2 \xrightarrow[H_2O]{Br_2} BrCH_2CH_2OH + HBr$$

Bromine water addition to an alkene

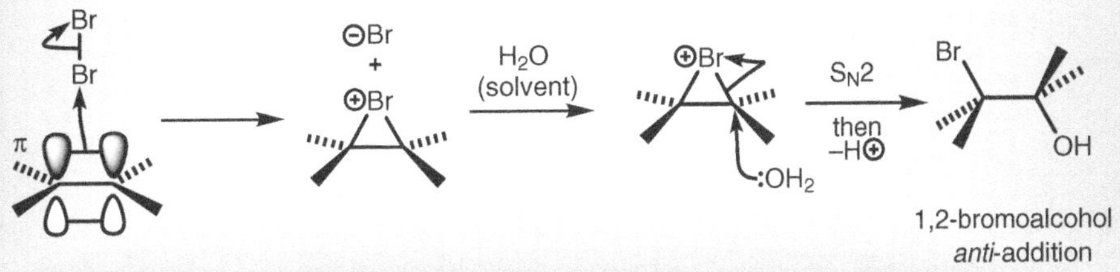
Br
Br
π
⊖Br
+
⊕Br
H2O
(solvent)
⊕Br
:OH2
SN2
then
–H⊕
Br
OH
1,2-bromoalcohol
anti-addition

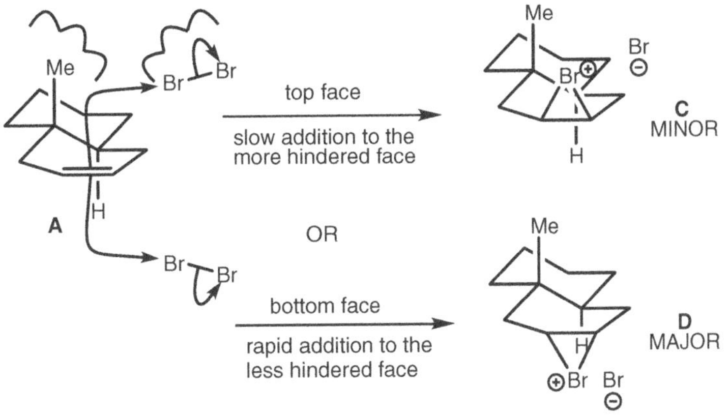
Me
Br
Br
H
A
top face
slow addition to the
more hindered face
Me
Br⊕
Br
⊖
H
C
MINOR
OR
Br
Br
bottom face
rapid addition to the
less hindered face
Me
H
⊕Br
Br
⊖
D
MAJOR

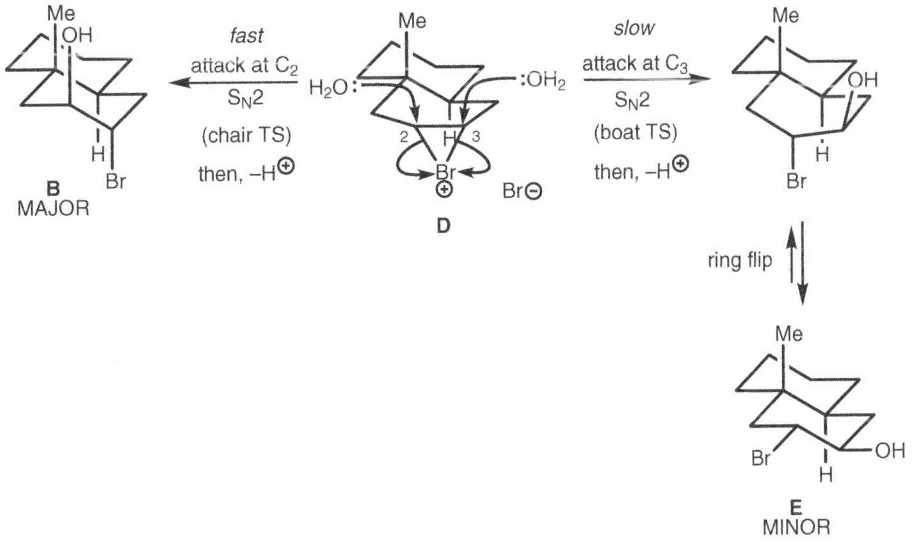
Me
OH
H
Br
B
MAJOR
fast
attack at C2
SN2
(chair TS)
then, –H⊕
Me
H2O:
:OH2
2
H
3
Br
⊕
Br⊖
D
slow
attack at C3
SN2
(boat TS)
then, –H⊕
Me
OH
H
Br
ring flip
Me
Br
H
OH
E
MINOR

Additional problems for Chapter 4

1. Addition of sodium hydroxide to a solution of cyclohexene **A** in chloroform gave the dichloride **B**. Draw a mechanism to account for this carbene reaction, and rationalise the stereochemical outcome.

Cyclohexene   NaOH / HCCl₃   bicyclic CCl₂ product

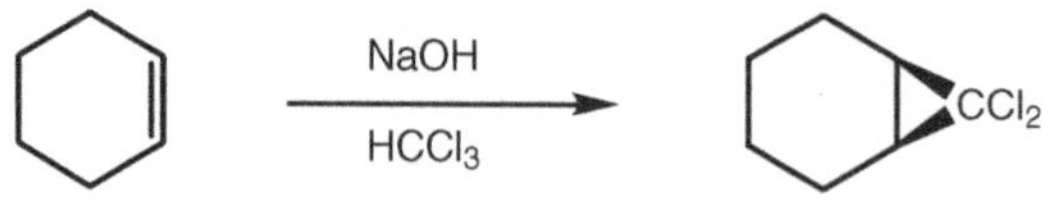

A   **B**

2. Suggest reagents for the following reaction. Account for the formation of 1,2-diol **D** from alkene **C**, and rationalise the stereochemical outcome.

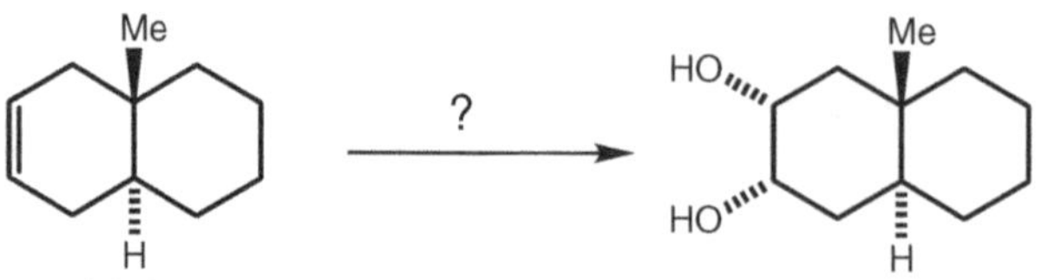

(±)-**C**   ?   (±)-**D**

3. Bromination of tiglic acid **E** gave a single vinyl bromide **F**. Draw a mechanism to account for this observation, and assign the configuration of the product.

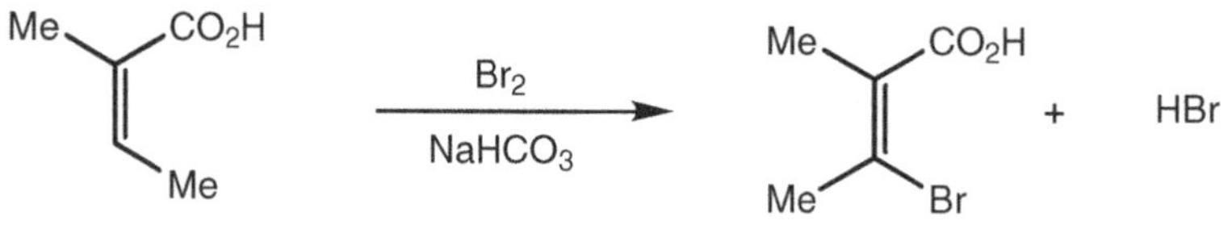

Me—CO₂H (with Me)   Br₂ / NaHCO₃   Me—CO₂H (Me, Br)   + HBr

E   **F**

4. The γ-lactone **H** can be formed directly by the addition of iodine to the alkenyl carboxylic acid **G**. Rationalise the stereochemistry present in the γ-lactone **H**, and draw a mechanism to account for this.

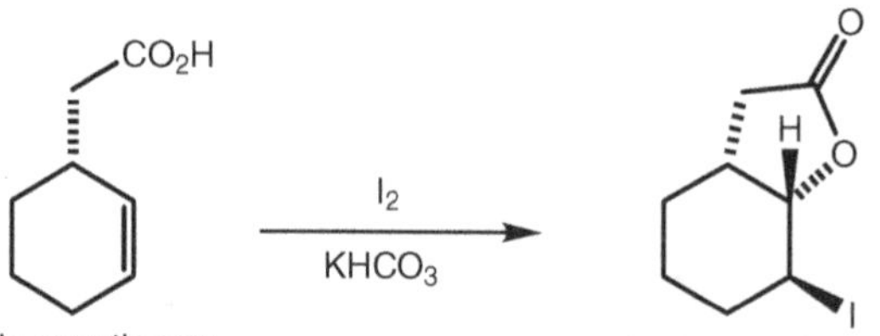

single enantiomer   I₂ / KHCO₃   single enantiomer

G   **H**

5. Hydroboration of the alkyne **I**, followed by protonolysis with acetic acid-d_1 (MeCO₂D) gave the styrene **J**, which contains a deuterium atom label. Rationalise the outcome of this reaction sequence and deduce the configuration present in the product **J**.

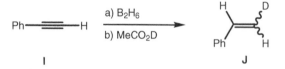

Ph—≡—H   a) B₂H₆   b) MeCO₂D   styrene product

I   **J**

Answers: 1. Addition of NaOH to HCCl₃ gives CCl₂. This dichlorocarbene adds stereospecifically *syn* to either face of the alkene **A** to give *meso*-dichloride **B**; 2. Formation of a *syn*-1,2-diol requires an OsO₄ mediated dihydroxylation; reagents OsO₄ and *t*-BuOOH (see page 80). OsO₄ adds to the less hindered face of alkene **C**, away from the adjacent axial bridge Me group (see page 90). Hydrolysis of the intermediate osmate ester gives the required 1,2-diol **D**; 3. *anti*-Addition of Br₂ across the C=C double bond, followed by stereospecific E2 elimination of HBr (H and Br need to be anti-periplanar) leads to (*Z*)-**F** (see pages 64 and 78); 4. This reaction is an iodolactonisation. I₂ addition to the less hindered face of the alkene (away from the CH₂CO₂H substituent) leads to the intermediate iodonium ion, which is intercepted by the tethered carboxylic acid group. Ring opening occurs on the nearer end (to give a five-membered lactone rather than a six), and on the opposite face of the iodonium ion because the process occurs via an intramolecular S_N2 reaction. This gives *anti*-stereochemistry between the C–I bond and the C–O bond in the *syn*-lactone; 5. B₂H₆ breaks down to give two equivalents of active BH₃. Hydroboration of the alkyne **I** gives a primary borane – B adds to less hindered carbon and H adds to the more hindered carbon (see page 88). Addition of acetic acid-d_1 allows a stereospecific BH₂-to-D interconversion to give (*E*)-**J** (see page 89 for the mechanism).

5 Addition to the Carbonyl Group

The aim of this chapter is to develop an understanding of the stereochemical features of the carbonyl group and its nucleophilic addition reactions.

At the end of this chapter, you should be able to:

- Draw and compare the mechanisms of nucleophilic additions to aldehydes and ketones.
- Predict and explain the use of facial selectivity.
- Explain the influence of neighbouring groups on the mechanisms and stereochemical outcome of these reactions.

Background

The carbonyl group of aldehydes and ketones is one of the more versatile functional groups in organic chemistry. Its behaviour is dominated by two very different types of reactions:

(1) Nucleophilic addition to an electron-deficient carbon atom of the carbonyl group, often accompanied by protonation of the oxygen atom.
(2) Electrophilic addition to electron-rich enols and enolates formed by deprotonation adjacent to the carbonyl group.

This chapter involves the first type; enols and enolates are considered in Chapter 6.

Nucleophilic addition to a carbonyl group

Addition to the carbon atom changes the planar, trigonal carbonyl group into a tetrahedral carbon atom. The reaction starts by donation of the nucleophile's non-bonded pair of electrons (its HOMO) into the unoccupied, antibonding orbital (LUMO, π^*) of the electrophilic C=O group. The direction of attack will be approximately 109° with respect to the trigonal plane (known as the Bürgi–Dunitz angle), which happens to be the bond angle present in the tetrahedral product.

The mechanism can be drawn in two dimensions:

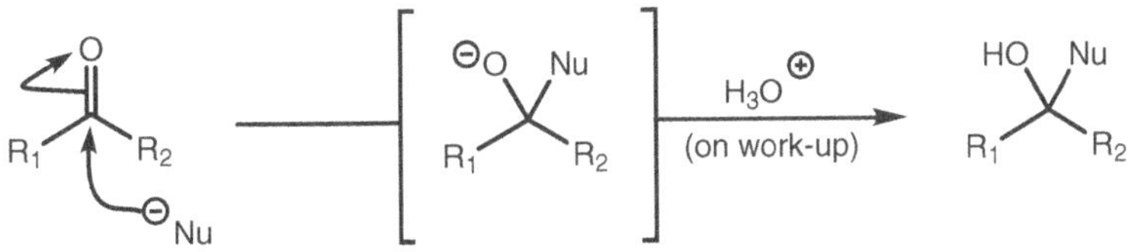

However, the Bürgi–Dunitz angle of nucleophilic attack is much easier to see when the mechanism is drawn in three dimensions.

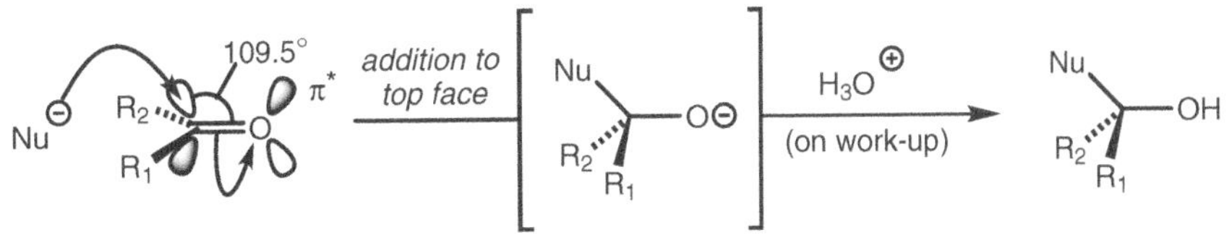

A new chiral centre can be introduced if the carbonyl compound is either an aldehyde (RCHO) or an unsymmetrical ketone (R_1R_2CO); the product will be racemic if both the reagent and substrate are achiral. If the two faces of the carbonyl plane are not identical (for example, when there is a chiral centre already present in the molecule), or if the reagent is itself asymmetric, then facial selectivity is possible.

Several factors will influence this facial selectivity. For reversible, equilibrating reactions, the major product will be the more stable one (thermodynamic control). For irreversible reactions, the major product will be the one that forms more readily (kinetic control); the chief considerations are: steric approach, product development, prior coordination of reagent with substrate and dipolar effects. The merits of each have to be judged for each reaction.

1. Draw all the products from the reaction of KCN in the presence of H_2SO_4 to the ketones **A** and **B**. Explain whether or not a new chiral centre is generated in this process.

single enantiomer

A   B

DRAW the STRUCTURES of A and B in three dimensions	Ketone **A** is achiral and ketone **B** is chiral. *Ketone **B** has C_2 symmetry and is non-superimposable on its mirror image; draw them or make models to check this.*
CONSIDER the REAGENTS and OUTLINE a MECHANISM	Cyanide ion is a good nucleophile. *HCN is a weak acid (pK_a 9, like a phenol) and so gives few cyanide ions at equilibrium. KCN is there to boost the concentration of cyanide ions.* Cyanide ion adds reversibly to ketones to give a tetrahedral, anionic intermediate which is then protonated to form the product, a **cyanohydrin**. The ⁻CN ion approaches the trigonal carbon atom from about 109° to the plane (Bürgi–Dunitz angle). *The non-bonded pair on the carbon atom in the cyanide ion is the HOMO, whereas the complementary LUMO is the π* of the C=O group.*
DRAW out a possible ANSWER and predict the major PRODUCT(S)	Cyanide ion could attack equally from either above or below the trigonal plane of ketone **A** or **B**. Each ketone gives only one cyanohydrin. *Seeing the single, chiral product from **B** is more difficult than seeing **A**'s. Use the mechanism to draw the two possible products from **B** and inspect them to see if they are the same or not. Notice that the new tetrahedral carbon atom has two identical substituents (as did the original ketone).*
SUMMARISE	Since the upper and lower faces of the ketones are the same, there is no facial selectivity and each ketone gives only one product. Due to the C_2 symmetry of **B**, addition to either top or bottom face leads to the same cyanohydrin **D**, in which the new tetrahedral carbon has two identical substituents. No new chiral centres have been generated in either **C** or **D**.

A

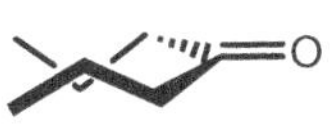

or

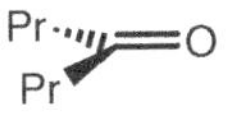
Pr
Pr
O

where Pr = $-CH_2CH_2CH_3$

B

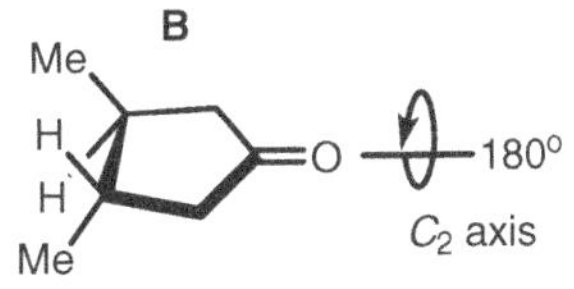
Me
H
H
Me
O
180°
C_2 axis

$$2 \times KCN + H_2SO_4 \longrightarrow 2 \times HCN + K_2SO_4$$

$$R_2C{=}O + HCN \longrightarrow R_2C(OH)CN$$

Nucleophilic addition of ^{-}CN followed by protonation

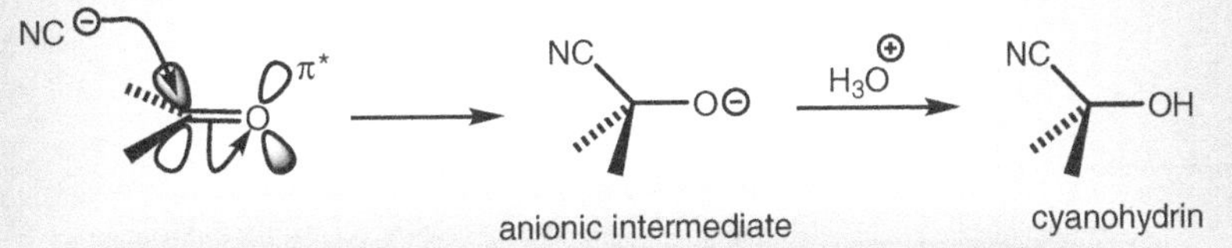
NC
π*
NC
O⊖
$H_3O^{\oplus}$
NC
OH
anionic intermediate
cyanohydrin

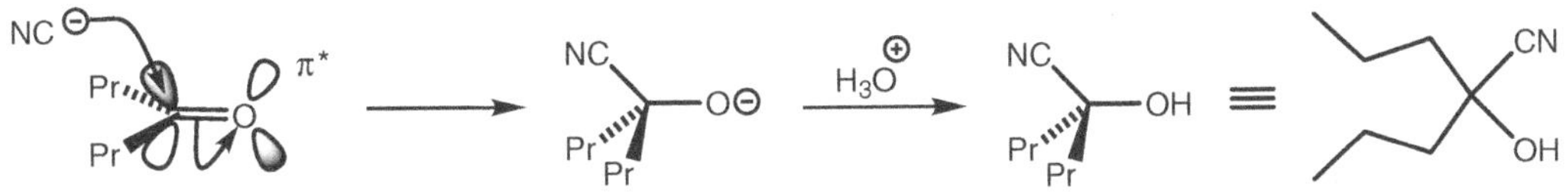
NC
Pr
Pr
π*
NC
O⊖
Pr
Pr
$H_3O^{\oplus}$
NC
OH
Pr
Pr
CN
OH

C

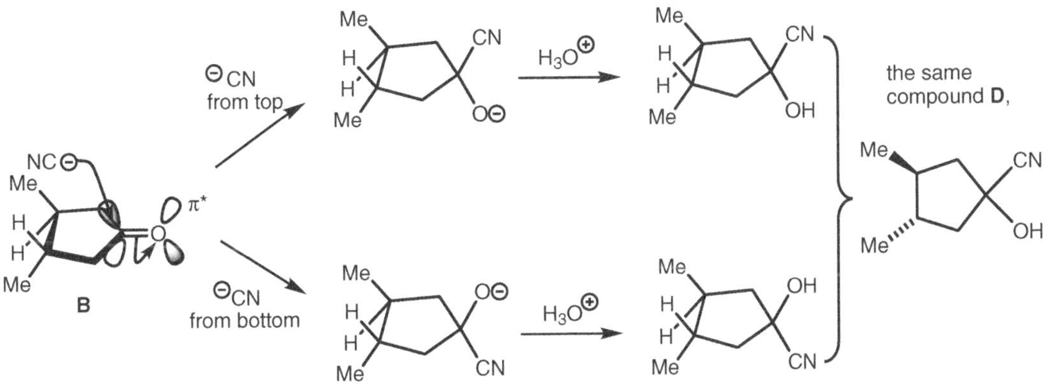
NC
Me
H
H
Me
π*
B
⊖CN
from top
⊖CN
from bottom
Me
H
H
Me
CN
O⊖
$H_3O^{\oplus}$
CN
OH
the same
compound **D**,
Me
Me
CN
OH

2. Predict the structure of the major product formed by reduction of the ketone **A** with aluminium prop-2-oxide in propan-2-ol. Rationalise the stereochemical outcome for this reaction.

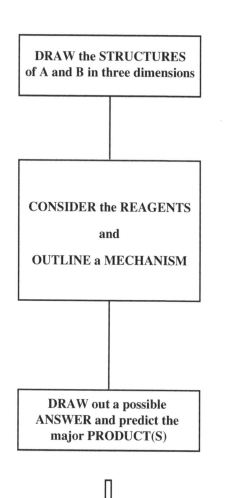

A

Al(OCHMe₂)₃
in Me₂CHOH

B

DRAW the STRUCTURES of A and B in three dimensions

These are both achiral structures. Each exists in a chair conformation with the large *tert*-butyl group in an equatorial position. The two faces of the ketone **A** are different (diastereotopic) because of the *tert*-butyl group. The two possible structures for the alcohol **B** are epimeric at C₁. *Ketone A and both isomers of alcohol B have a plane of symmetry.*

CONSIDER the REAGENTS and OUTLINE a MECHANISM

Al(OCHMe₂)₃ in Me₂CHOH reduces ketones to secondary alcohols, the propan-2-ol being oxidised at the same time to propanone (this is the **Meerwein–Ponndorf reduction**).

The aluminium alkoxide coordinates to the ketone's carbonyl oxygen atom. A hydride ion (a hydrogen atom with its bonding electrons) is then transferred to the C=O via a six-membered transition state. The reaction is reversible, so the most stable compound will predominate at equilibrium (thermodynamic control), and *not* the one that is most easily formed (kinetic control).
Attack is from the Bürgi–Dunitz angle.
HOMO: CH bonding pair; LUMO: C=O π.*

DRAW out a possible ANSWER and predict the major PRODUCT(S)

Attack from the *axial* direction (the top as drawn) gives the more stable alkoxide *anti*-**C** which in turn leads to the more stable alcohol *anti*-**B** through proton exchange with Me₂CHOH. Attack from the *equatorial* direction (below) gives the less stable alkoxide *syn*-**C** and alcohol *syn*-**B**. Both *syn*-**C** and *syn*-**B** suffer from steric crowding by the axial hydrogens across the ring.
Making the original aluminium complex with A may be easier from below, but this is irrelevant when the reaction is at equilibrium.

SUMMARISE

The faces of ketone **A** are diastereotopic and so facial selectivity is observed. Meerwein–Ponndorf reduction is reversible and occurs by hydride ion transfer to the top face of the ketone via a six-membered ring transition state to give the more stable, less hindered diastereoisomeric secondary alcohol, *anti*-**B**.

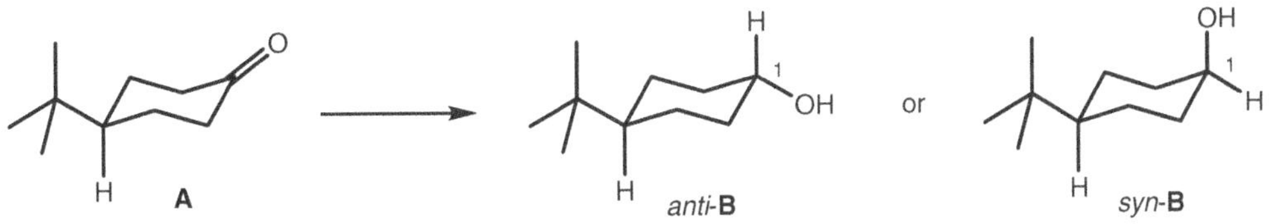
O
H
1
OH
or
OH
1
H
H
A
H
anti-B
H
syn-B

Plane of symmetry in **A** (passes through C=O, C_4, H_4 and C_7)

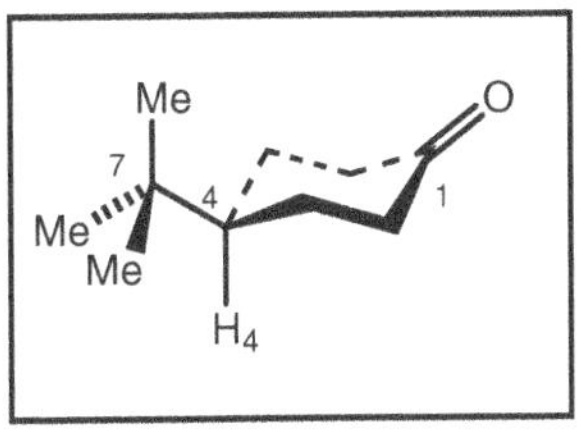
Me
7
Me
Me
4
O
1
H_4

$$R_2C{=}O + Me_2CHOH \underset{}{\overset{Al^{3+}}{\rightleftharpoons}} R_2CHOH + Me_2C{=}O$$

Meerwein–Ponndorf reduction: hydride transfer

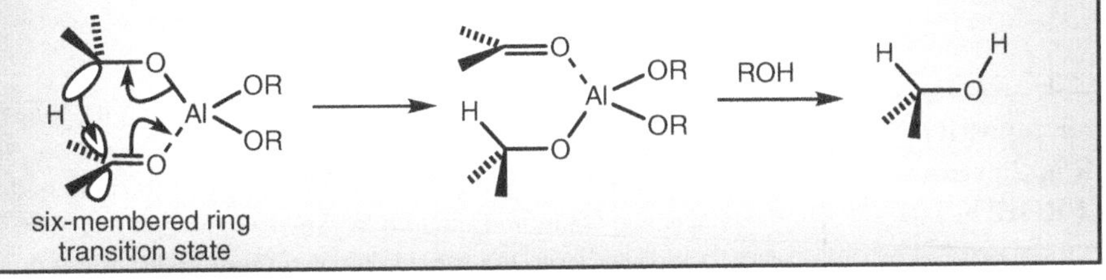
O
H
Al
OR
OR
O
O
Al
OR
OR
H
O
ROH
H
H
O
six-membered ring
transition state

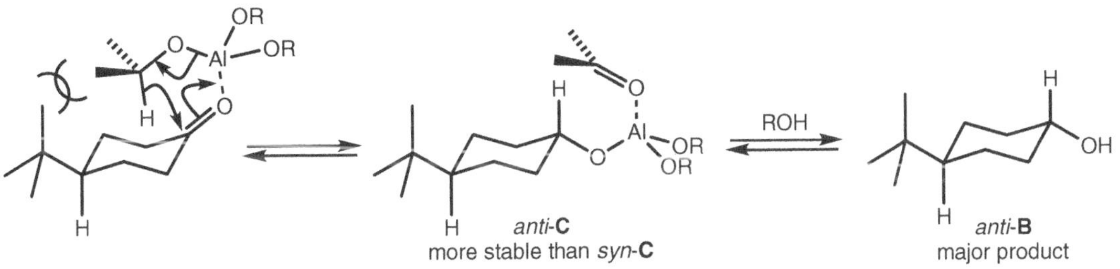
OR
O
Al
OR
H
O
H
H
O
Al
OR
OR
O
ROH
H
OH
H
anti-C
more stable than syn-C
H
anti-B
major product

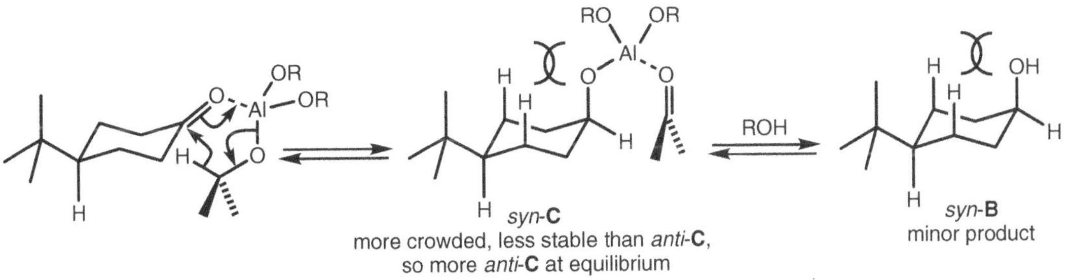
OR
O
Al
OR
H
O
H
RO
OR
Al
O
O
H
H
H
ROH
H
OH
H
H
H
syn-C
more crowded, less stable than anti-C,
so more anti-C at equilibrium
H
syn-B
minor product

3. Account for the difference in selectivity upon reduction of the bicyclic ketones **A** and **B** with NaBH₄ in MeOH.

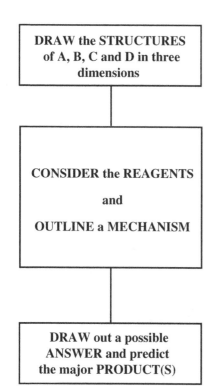

DRAW the STRUCTURES of A, B, C and D in three dimensions	Both the ketones **A** and **B** are chiral; each has two chiral centres. The conformations are fixed due to the rigidity of the bicyclic framework. The product alcohols are also chiral and a new chiral centre is formed in each reaction.

CONSIDER the REAGENTS

and

OUTLINE a MECHANISM

NaBH₄ is a nucleophilic hydride-ion donor which reduces ketones to secondary alcohols.
NaBH₄ is not a source of free hydride ions, H⁻, but donates its hydride by using its nucleophilic ⁻B–H bond directly.

Borohydride reduction is not reversible. The BH₄⁻ ion approaches the trigonal carbon of C=O at the Bürgi–Dunitz angle, 109° to the plane. *For an irreversible reaction, consider the stabilities of the transition states leading to the possible products (compare this kinetically controlled reaction with the one in question (2) (on page 96) in which the reaction is reversible and under thermodynamic control).*
HOMO: BH bonding pair; LUMO: C=O π*.

DRAW out a possible ANSWER and predict the major PRODUCT(S)

From the structures of *endo*-**C** and *exo*-**D**, BH₄⁻ must have approached **A** from the top face and **B** from the lower face. Access to the lower face of **A** or **B** is restricted by the *endo* H atoms on the underside. For **A**, topside approach is more favoured leading to the alcohol *endo*-**C**. However, for **B** approach from the topside is even more disfavoured due to the presence of the CH₃ group on the neighbouring bridge position, so approach comes from below, slowly, giving *exo*-**D**.
*Making a model of **A** or **B** will show you that the C=O points slightly downwards, so that Bürgi–Dunitz approach from below is very close to the endo CH bonds. Approach from above is angled away from the bulk of the molecule, so that the bridge H atom in **A** does not interfere.*

SUMMARISE

Reduction occurs from the less hindered face of each ketone. Because both ketones are chiral, the faces of the carbonyl groups are different (diastereotopic) and thus addition rates will be unequal. A single diastereoisomeric secondary alcohol is formed from each due to steric approach control.

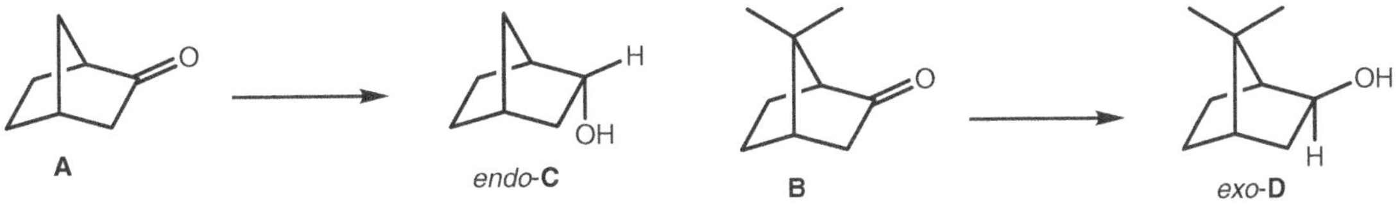

$$R_2C=O \xrightarrow[\text{MeOH}]{\text{NaBH}_4} R_2CHOH$$

Reduction of ketone to secondary alcohol

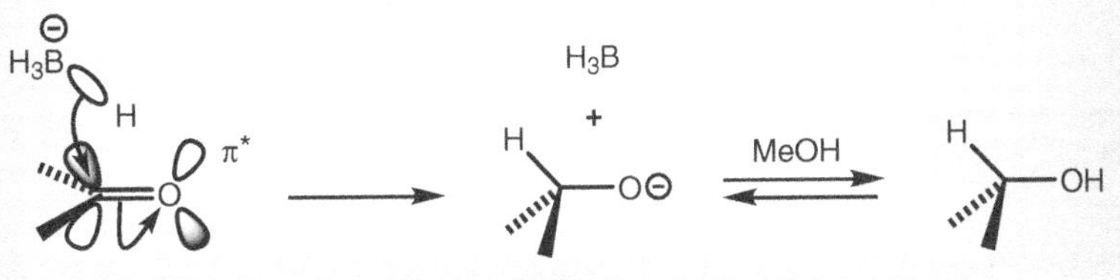

'Hydride' addition to C=O π^* at 109° to the plane

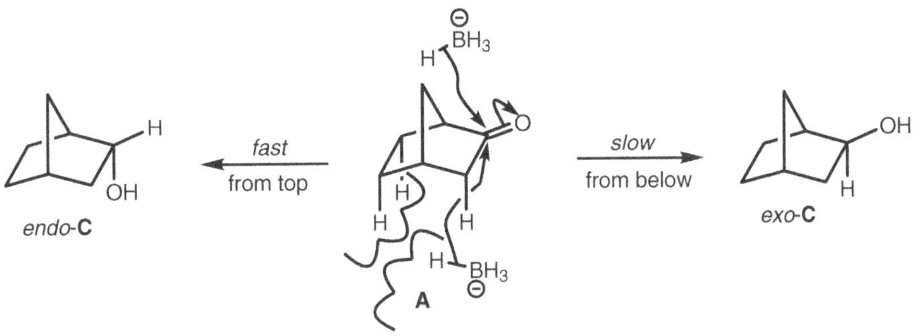

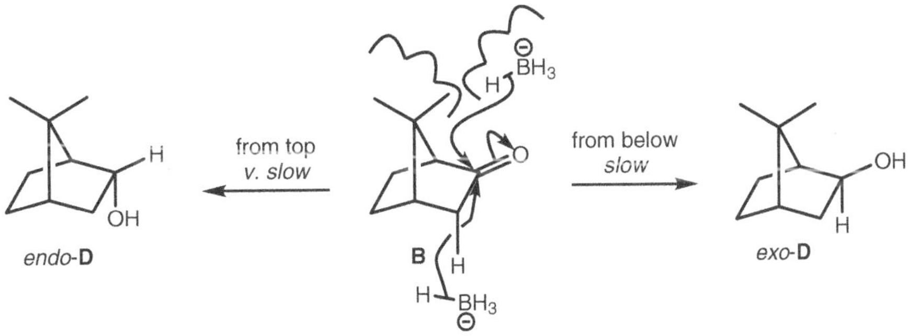

So:

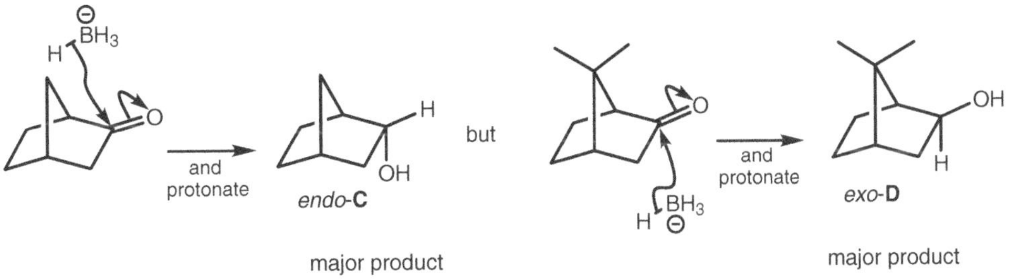

4. Reduction of the ketone **A** with L-Selectride™, LiBH[CH(Me)Et]₃ gave a diastereoisomeric mixture (ratio 93:7) of the alcohols *syn*- and *anti*-**B**. Account for this observation.

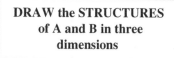

DRAW the STRUCTURES of A and B in three dimensions	Ketone **A** and the two secondary alcohols are all achiral. The faces of the ketone are diastereotopic. Alcohols *syn*-**B** and *anti*-**B** are epimers. *Draw out L-Selectride in full.*
CONSIDER the REAGENTS and OUTLINE a MECHANISM	LiBH[CH(Me)Et]₃ is a nucleophilic hydride-ion source similar to NaBH₄, but much bulkier and more reactive. *Here we have a relatively unhindered ketone and a hindered reagent, the opposite way around from question (3) on page 98.* This reduction is not reversible; therefore the reaction is under kinetic control. The L-Selectride will attack at the Bürgi–Dunitz angle; note also the large size of the three CH(Me)Et groups. *Approach path and transition state stability will be important.* *HOMO: BH bonding pair; LUMO: C=O π*.*
DRAW out a possible ANSWER and predict the major PRODUCT(S)	From the structure of *syn*-**B**, the reagent must deliver the hydride ion mostly to the lower face of the ketone **A**. For this bulky reagent, access across the top face of **A** (axial approach) to the carbonyl carbon is severely restricted by the interaction between the large CH(Me)Et groups on the boron and the vertical, axial H atoms on the chair conformer. For equatorial approach to the lower face, the ideal approach angle gives more space for the CH(Me)Et groups so that the BH bond can get closer access to the C=O. *Remember that the C=O in A points slightly upwards, so that Bürgi–Dunitz approach from below is angled **away** from the bulk of the molecule, so the axial H atoms on the underneath of A do not get in the way.*
SUMMARISE	Ketone **A** has diastereotopic faces. The reaction is under kinetic control and approach of the bulky reducing agent is facially selective, giving *syn*-**B** as the major product. Axial approach giving *anti*-**B** is hindered by interaction with the axial hydrogen atoms.

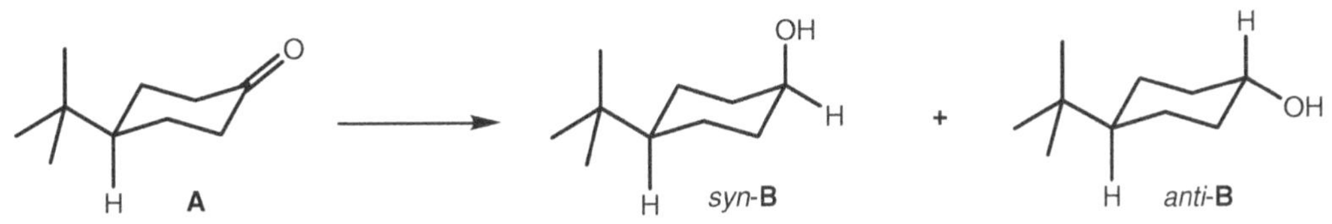
O
OH
H
+
H
OH
H
A
H
syn-B
H
anti-B

L-Selectride: $LiHB[CH(Me)CH_2Me]_3$ ≡

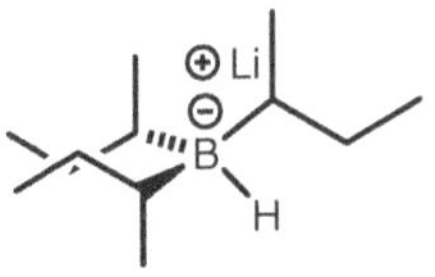
⊕ Li
⊖
B
H

$$R_2C{=}O \xrightarrow[\text{then MeOH}]{LiHBR_3} R_2CHOH$$

Reduction of ketone to secondary alcohol

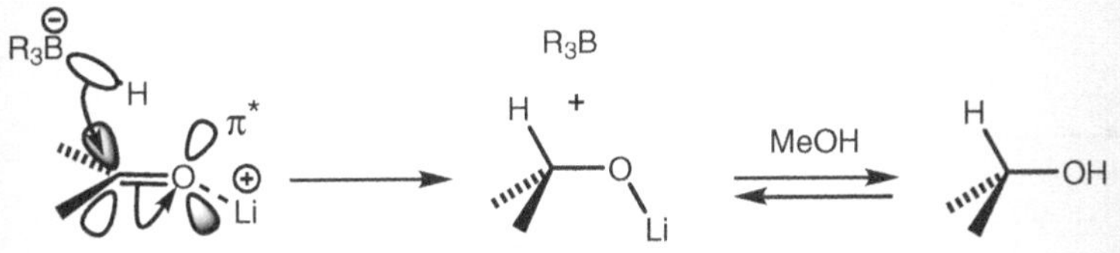
⊖
R₃B
H
π*
O
⊕
Li
R₃B
H +
O
Li
MeOH
H
OH

'Hydride' addition to C=O π* at 109° to the plane

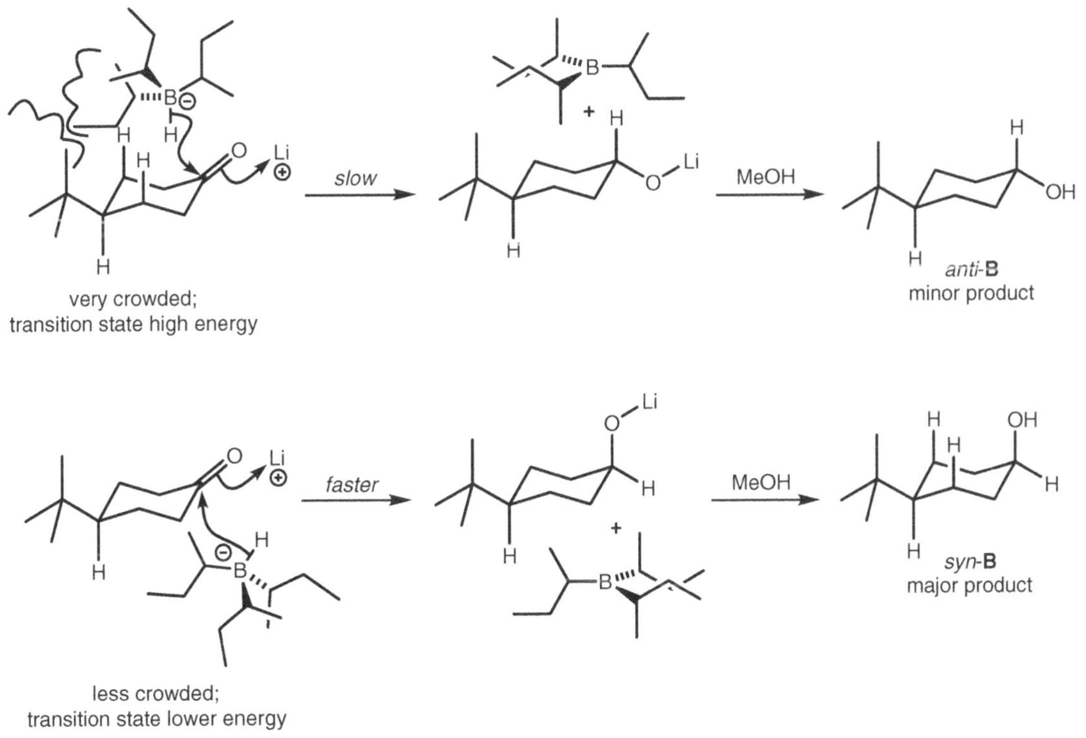
B⊖
H
H
H
O
Li
⊕
H
slow
B
+ H
O
Li
H
MeOH
H
OH
H
anti-B
minor product
very crowded;
transition state high energy
O
Li
⊕
faster
H
⊖
B
H
O
Li
H
H
+
B
MeOH
H
OH
H
H
H
syn-B
major product
less crowded;
transition state lower energy

5. Rationalise the stereoselectivity observed by reduction of the chiral ketone (*R*)-**A** with LiAlH₄.

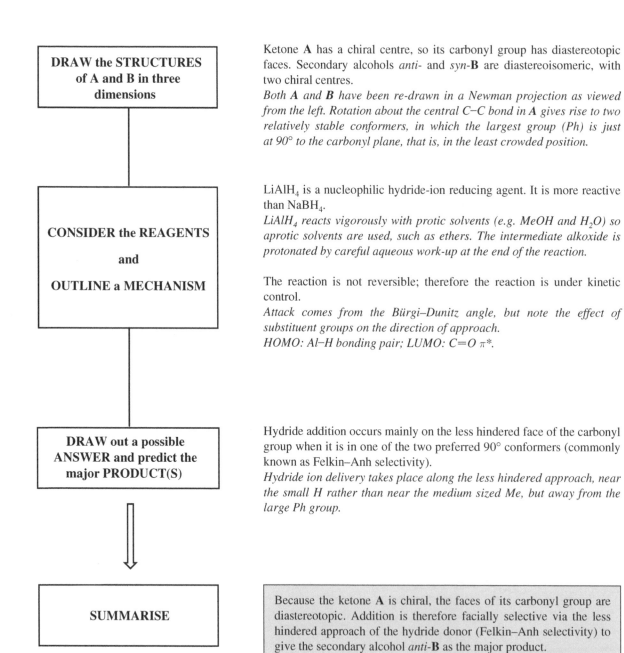

DRAW the STRUCTURES of A and B in three dimensions

Ketone **A** has a chiral centre, so its carbonyl group has diastereotopic faces. Secondary alcohols *anti*- and *syn*-**B** are diastereoisomeric, with two chiral centres.

*Both **A** and **B** have been re-drawn in a Newman projection as viewed from the left. Rotation about the central C–C bond in **A** gives rise to two relatively stable conformers, in which the largest group (Ph) is just at 90° to the carbonyl plane, that is, in the least crowded position.*

CONSIDER the REAGENTS

and

OUTLINE a MECHANISM

LiAlH₄ is a nucleophilic hydride-ion reducing agent. It is more reactive than NaBH₄.

LiAlH₄ reacts vigorously with protic solvents (e.g. MeOH and H₂O) so aprotic solvents are used, such as ethers. The intermediate alkoxide is protonated by careful aqueous work-up at the end of the reaction.

The reaction is not reversible; therefore the reaction is under kinetic control.

Attack comes from the Bürgi–Dunitz angle, but note the effect of substituent groups on the direction of approach.
HOMO: Al–H bonding pair; LUMO: C=O π*.

DRAW out a possible ANSWER and predict the major PRODUCT(S)

Hydride addition occurs mainly on the less hindered face of the carbonyl group when it is in one of the two preferred 90° conformers (commonly known as Felkin–Anh selectivity).

Hydride ion delivery takes place along the less hindered approach, near the small H rather than near the medium sized Me, but away from the large Ph group.

SUMMARISE

Because the ketone **A** is chiral, the faces of its carbonyl group are diastereotopic. Addition is therefore facially selective via the less hindered approach of the hydride donor (Felkin–Anh selectivity) to give the secondary alcohol *anti*-**B** as the major product.

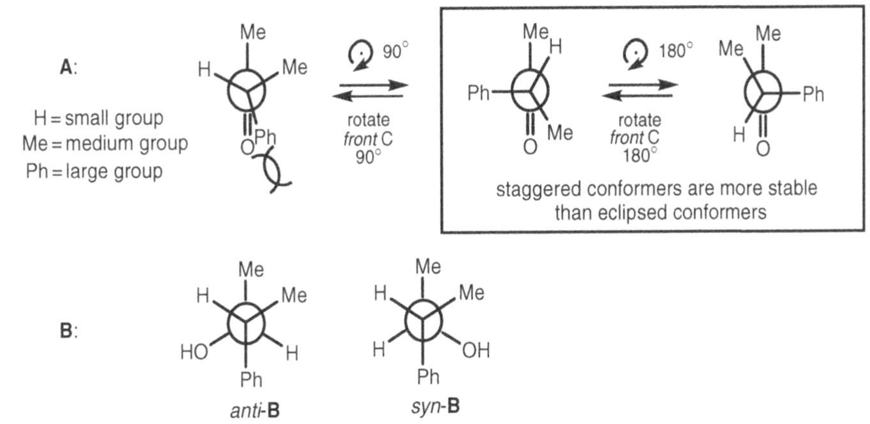
A:
H = small group
Me = medium group
Ph = large group
rotate
front C
90°
rotate
front C
180°
staggered conformers are more stable
than eclipsed conformers
B:
anti-B
syn-B

$$R_2C{=}O \xrightarrow[\text{then } H_3O^{\oplus}]{LiAlH_4} R_2CHOH$$

Reduction of ketone to secondary alcohol

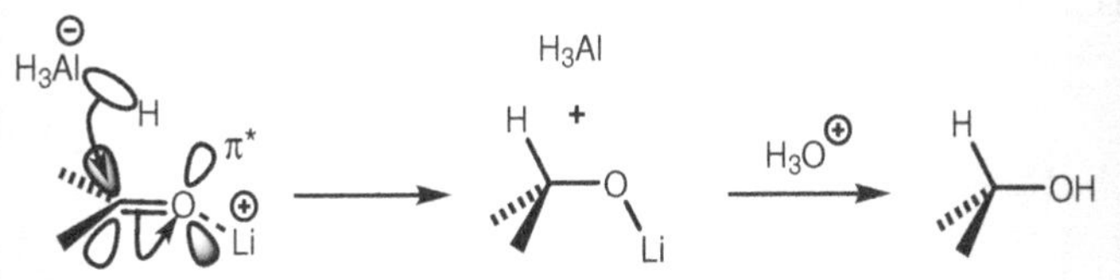

'Hydride' addition to C=O π* at 109° to the plane

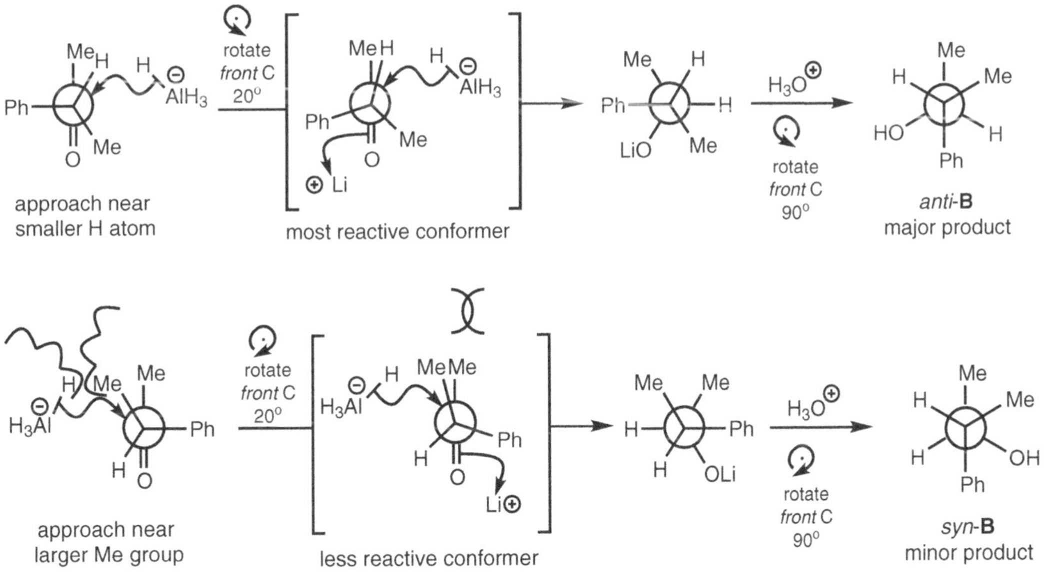
approach near
smaller H atom
rotate
front C
20°
most reactive conformer
rotate
front C
90°
anti-B
major product
approach near
larger Me group
rotate
front C
20°
less reactive conformer
rotate
front C
90°
syn-B
minor product

6. Addition of excess PhLi to the α-hydroxy ketone (S)-**A** gave predominantly the diol (S,S)-**B**. Rationalise the stereo-chemical outcome of this reaction.

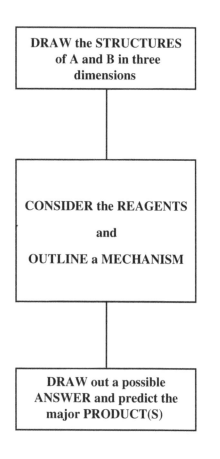

| DRAW the STRUCTURES of A and B in three dimensions | The α-hydroxy ketone (S)-**A** is chiral with diastereotopic faces. Hydrogen bonding is possible between the C=O and OH groups. The secondary alcohols, (S,S)- and (S,R)-**B** are diastereoisomeric, with two chiral centres each: (S,S)-**B** is chiral, (S,R)-**B** is *meso-*. |

*Only one conformer of **A** has an internal hydrogen bond. (Compare with questions (4) and (5) on pages 100 and 102 respectively, in which the ketones have no intramolecular hydrogen bonding.)*

CONSIDER the REAGENTS

and

OUTLINE a MECHANISM

PhLi is an organometallic reagent in which the C-metal bond is polarised $Ph^{\delta-}$-$Li^{\delta+}$. PhLi is both basic and nucleophilic, and will add to ketones to give tertiary alcohols. Lithium ion can coordinate to electron-donor groups.

First the OH group is deprotonated and then irreversible addition to the carbonyl takes place under kinetic control.

As the pK_a of ROH is about 16 and the pK_a of benzene is 43, PhLi will deprotonate ROH to give PhH and ROLi. Again attack will come from the Bürgi–Dunitz angle.
HOMO: Metal-H bonding pair; LUMO: C=O π^*.

DRAW out a possible ANSWER and predict the major PRODUCT(S)

A planar, five-membered ring lithium chelate is formed from **A** using the lone pairs on the oxygen atoms of the C=O and O^- groups. A second PhLi (perhaps coordinated to the oxygens) then delivers Ph^- faster to the less hindered face of the carbonyl group.

Approach past the smaller Me is less hindered than the route past the bigger Ph group. If the incoming PhLi is complexed to the alkoxides $-O^-$, then this effect may be more pronounced.

SUMMARISE

The chiral α-hydroxy ketone forms a diastereotopic lithium chelate. Nucleophilic addition of PhLi selectivity to its less hindered face gives the chiral tertiary alcohol (S,S)-**B** as the major product.

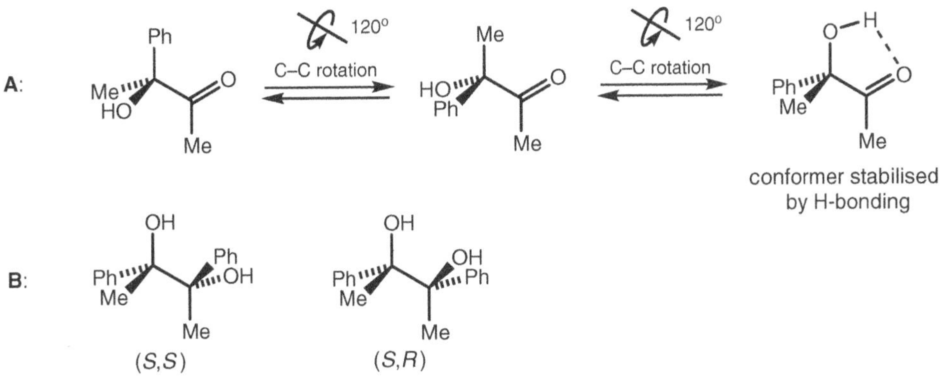
A:
Ph
Me
HO
O
Me
120°
C–C rotation
Me
HO
Ph
O
Me
120°
C–C rotation
O–H
O
Ph
Me
Me
conformer stabilised by H-bonding
B:
OH
Ph
Ph
OH
Me
Me
(S,S)
OH
OH
Ph
Ph
Me
Me
(S,R)

$$ROH + R'Li \longrightarrow ROLi + R'H \xrightarrow{H_3O^{\oplus}} ROH$$

$$R_2C{=}O + R'Li \longrightarrow R_2R'COLi \xrightarrow{H_3O^{\oplus}} R_2R'COH$$

Nucleophilic addition of R′Li, followed by protonation

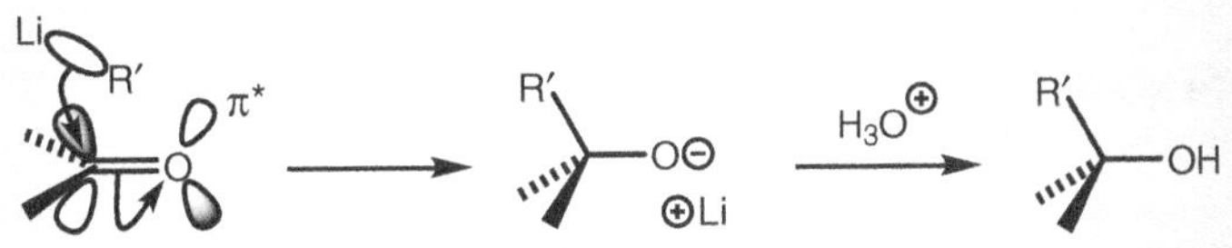
Li
R′
π*
O
R′
O⊖
⊕Li
H3O⊕
R′
OH

R′Li addition to C=O π* at 109° to the plane

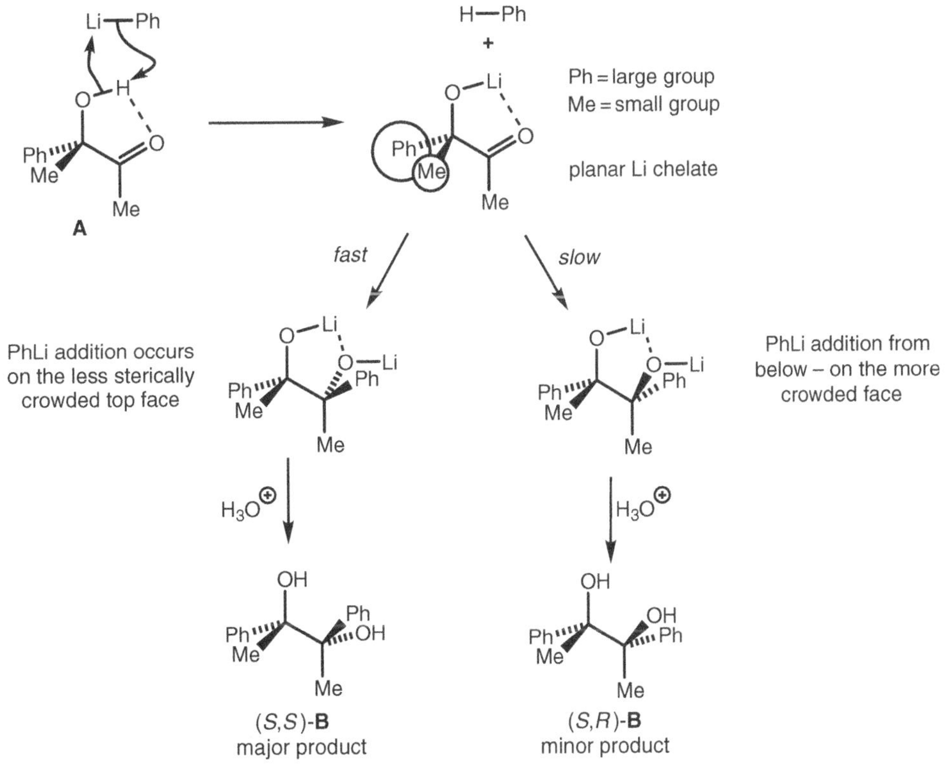
Li–Ph
O–H
O
Ph
Me
Me
A
H–Ph
+
O–Li
O
Ph
Me
Me
Ph = large group
Me = small group
planar Li chelate
fast
slow
PhLi addition occurs on the less sterically crowded top face
O–Li
O–Li
Ph
Ph
Me
Me
O–Li
O–Li
Ph
Ph
Me
Me
PhLi addition from below – on the more crowded face
H3O⊕
H3O⊕
OH
Ph
Ph
OH
Me
Me
(S,S)-B
major product
OH
OH
Ph
Ph
Me
Me
(S,R)-B
minor product

7.   Predict the structure of the major product in the following reaction. Explain your answer.

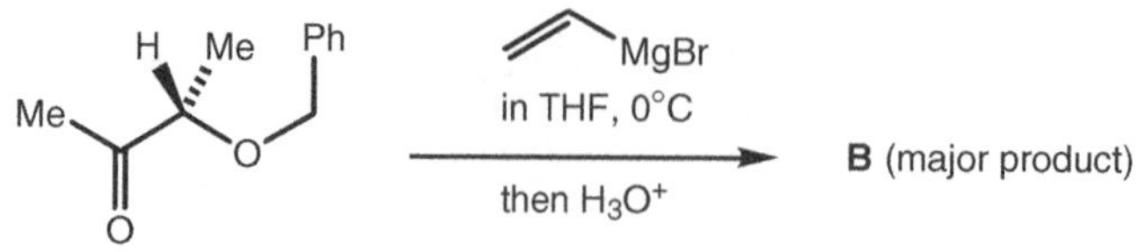

B (major product)

(S)-**A**

DRAW the STRUCTURE of A in three dimensions	The α-alkoxy ketone **A** is chiral with diastereotopic faces. Its single chiral centre has an (S)-configuration. *Either perspective or Newman projections might be useful. No internal hydrogen bond here (compared with questions (5) and (6) on pages 102 and 104 respectively, but note the presence of the ether group).*
CONSIDER the REAGENTS **and** **OUTLINE a MECHANISM**	CH_2=CHMgBr is a Grignard reagent in which the *C*-metal bond is polarised: CH_2=CH$^{\delta-}$-Mg$^{\delta+}$. Grignard reagents are both basic and nucleophilic, and will add to ketones to give tertiary alcohols. The magnesium ion can coordinate to electron-donor groups. Irreversible addition to carbonyl takes place under kinetic control. *Attack will come from the Bürgi–Dunitz angle as usual.* *HOMO: Metal-R bonding pair; LUMO: C=O π*.*
DRAW out a possible ANSWER and predict the major PRODUCT(S)	A planar, five-membered ring magnesium chelate can be formed from **A** using the oxygen lone pairs of the C=O and OCH$_2$Ph groups as donors. This Grignard reagent (or a second CH_2=CHMgBr coordinated to the first) then delivers the ethenyl group faster to the less hindered face of the carbonyl group. *Approach on the side of the smaller H is less hindered and more accessible than past the larger Me group.*
SUMMARISE	The chiral α-alkoxyketone forms a diastereotopic magnesium chelate. Nucleophilic addition of CH_2=CHMgBr selectivity to the less hindered face of the chelate gives the chiral tertiary alcohol (S,S)-**B** as the major product.

A:

or,

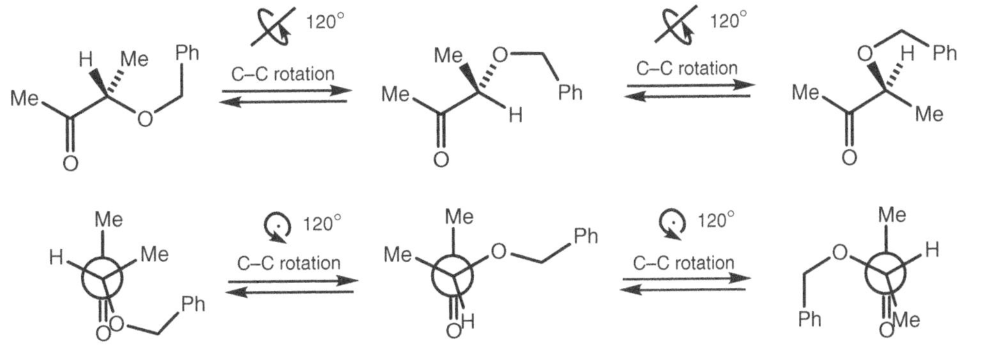
120°
C–C rotation
Me
H
Ph
O

$$R_2C{=}O \ + \ R'MgBr \longrightarrow [R_2R'COMgBr] \xrightarrow{H_3O^{\oplus}} R_2R'COH$$

Nucleophilic addition of RMgBr, followed by protonation

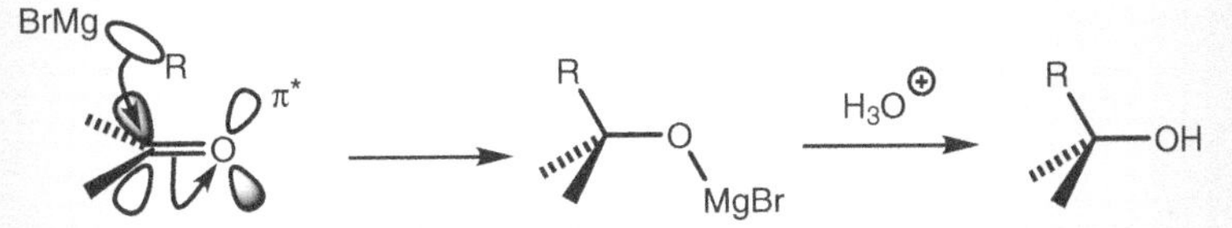
BrMg
R
π*
O
MgBr
H3O⊕
OH

Grignard addition to C=O π^* at 109° to the plane

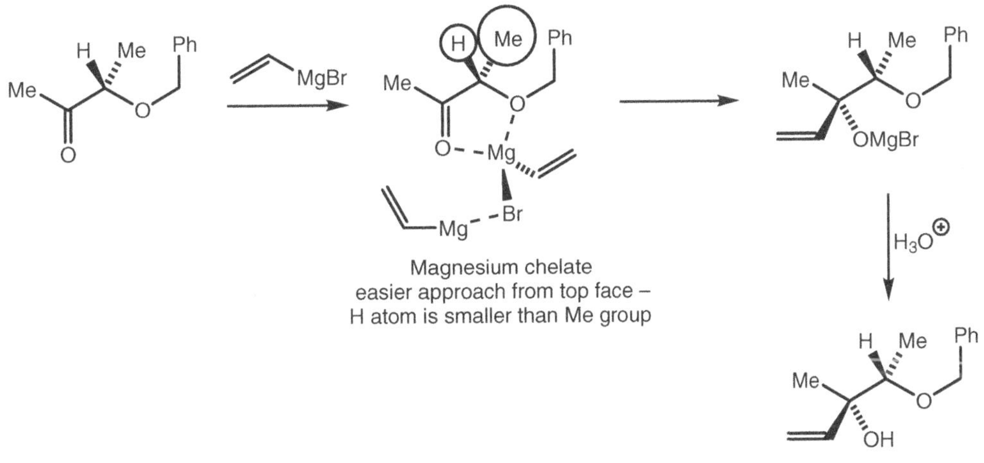
MgBr
H
Me
Ph
O
Mg
Br
OMgBr
H3O⊕
OH
Magnesium chelate
easier approach from top face –
H atom is smaller than Me group

major product

Assign new stereogenic centre:

Four groups	CH_3	$CHCH_2$	OH	$CHMeOCH_2Ph$
First atom	C	C	**O**	C
Second atom	**H**,H,H	**C**,C,H		**O**,C,H
Priorities	4	3	1	2

Product **B** has (*S*,*S*)-stereochemistry

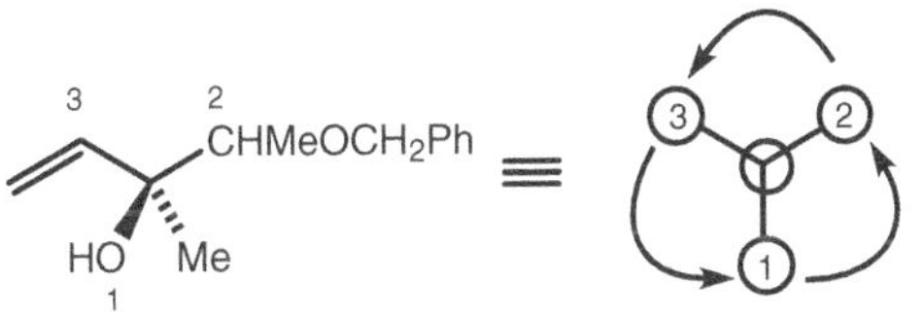
3
2
CHMeOCH2Ph
HO
Me
1

anticlockwise, so (*S*)-

Additional problems for Chapter 5

Section A

Draw diagrams to show the two isomeric products formed in each of the following reactions:

1. **A** $\xrightarrow{\text{NaBH}_4 \text{ in EtOH}}$

2. **B** $\xrightarrow{\text{NaC}\equiv\text{CH, then H}_3\text{O}^+}$

3. **C** $\xrightarrow{\text{KCN + H}_2\text{SO}_4}$

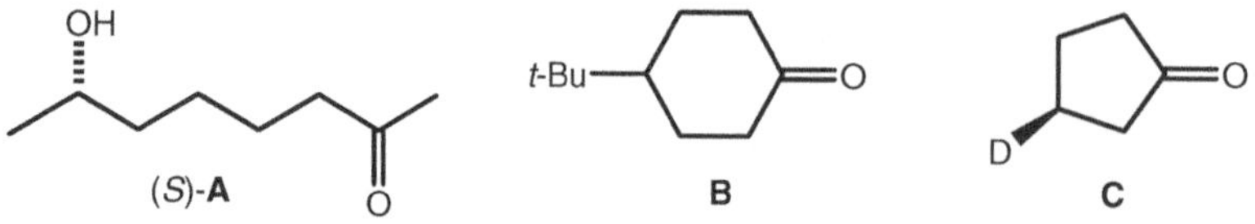

(S)-**A**   **B**   **C**

For each product, state whether it is chiral or not.

Section B

Account for the following reactions.

1.

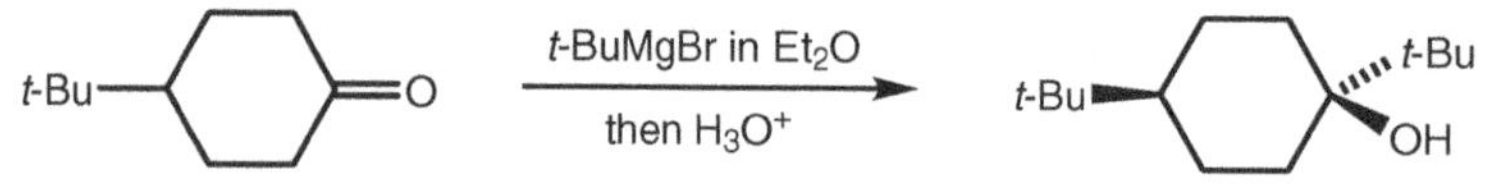

2.

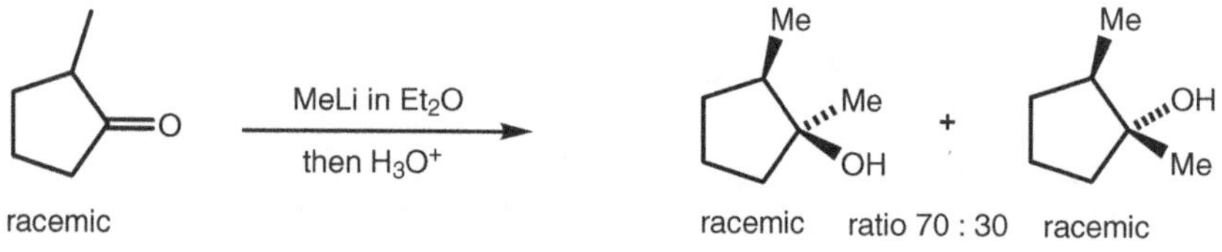

racemic   racemic   ratio 70 : 30   racemic

3.

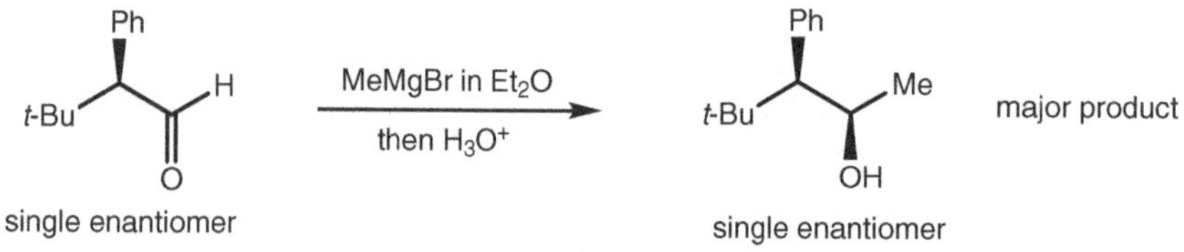

single enantiomer   single enantiomer

4.

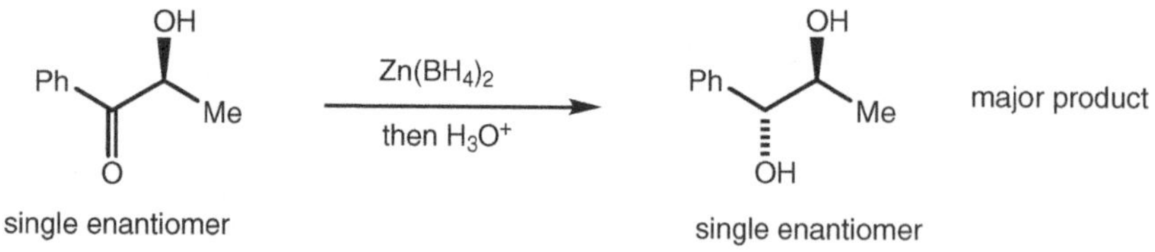

single enantiomer   single enantiomer

5.

single enantiomer   single enantiomer

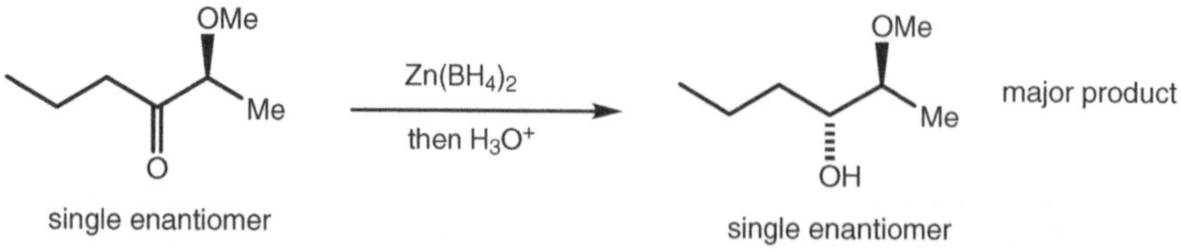

Answers: A1. Chiral (*S,S*) + achiral (*meso-*) (*S,R*)-diols; A2. *Syn-* and *anti-*adducts, both achiral (plane of symmetry); A3. *Syn-* and *anti-*adducts, both chiral; B1. Large reagent less crowded equatorial approach (see page 100); B2. Crowded substrate, easier approach to the face away from the Me group (see page 98); B3. Felkin-Anh: large group (*t*-Bu) at 90° to C=O (see page 102); B4. Chelation of Zn ion by OH and C=O (see page 104); B5. Chelation of Zn ion by OMe and C=O (see page 106).

6 Chemistry of Enolates

The aim of this chapter is to develop an understanding of the stereochemical features of enolates and their electrophilic addition reactions.

At the end of this chapter, you should be able to:

- Draw and describe the configuration of a given enolate.
- Explain the stereochemical differences between reactions involving kinetic and thermodynamic enolate formation.
- Draw stereo- and the mechanisms for electrophilic addition to an enolate.
- Rationalise any facial selectivity.

Background

Enols and enolates are very important intermediates in organic and biological chemistry. The typical reaction of a carbonyl group involves nucleophilic addition to the carbon atom of the C=O bond. In contrast, enols and enolates are electron-rich species and react with electron-deficient species. The double bond of the enolate can be thought of as an oxy-substituted alkene and can show the same (E)- and (Z)-stereochemistry or facial selectivity as an alkene. Initial addition of an electrophile E^+ to an enolate or an alkene is somewhat similar, but for an enolate the second stage usually involves re-formation of the stable C=O bond rather than addition of the counter-ion X^-.

- tetrahedral carbon, sp^3 hybridised   ⊗ trigonal planar carbon, sp^2 hybridised

Formation and reactions of enolates

Deprotonation at the sp^3 hybridised tetrahedral C−H bond next to the C=O bond is more efficient when the orientation of the C−H bond allows good overlap between the bonding electrons of the C−H σ-bond (HOMO) and the antibonding π*-orbital of the C=O bond (LUMO). The resulting sp^2 hybridised enolate is a planar anion, delocalised by involvement of the electron pairs in the π-bond with a non-bonded pair on oxygen.

Enolate formation can be either irreversible (kinetic control) or reversible (thermodynamic control). The main influences on the stereochemistry of the resulting enolate include: (a) the most acidic H lying approximately perpendicular to the C=O plane is removed; (b) large groups are kept apart and (c) metal counterions are stabilised by coordination. Subsequent electrophilic addition to the enolate occurs from above (or below) the C=C plane, where the HOMO of this delocalised 4-electron, 3-atom π-system has its highest electon density.

For example, base-catalysed bromination of a ketone results in the formation of a new C−Br bond by re-formation of the more stable C=O bond.

1. The lithium enolate **Y** can be be formed reversibly from ethyl 3-oxo-butanoate **A**. Predict and rationalise the structure and stereochemistry of **Y**.

Ethyl 3-oxo-butanoate, **A** $\xrightarrow{\text{LiOR}}$ **Y** ($C_6H_9O_3Li$)

DRAW the STRUCTURE of ethyl 3-oxo-butanoate	**A** is a β-ketoester and has no chirality. *A is commonly known as ethyl acetoacetate. Looking at the molecular formulae of A and Y, one H is lost and Li is gained in the reaction.*

CONSIDER the REAGENTS and OUTLINE a MECHANISM

Lithium alkoxides are basic. Deprotonation of the C–H adjacent to a C=O group gives a resonance stabilised enolate which may be (*E*)- or (*Z*)- across the new C=C bond. Reversibility suggests that the more stable enolate will be formed.

DRAW out a possible ANSWER and predict the major PRODUCT(S)

For **A**: deprotonation at C_2 gives an anion in which the charge is delocalised over both adjacent C=O groups. **D** is the biggest contributor (negative charge on an electronegative atom, ester resonance unimpaired). Its stereoisomer **D-2** is less crowded than **D-1**. Drawing **D-2a** shows **D-2** with the additional coordination of Li to the carbonyl oxygen of the ester group; this stabilising interaction cannot be present in **D-1**. **D-2a** has a (*Z*)-configuration and is the most stable enolate from **A**.
The hydrogens on C_2 are more acidic than those on C_4. β-Keto esters (pK_a about 10) are more acidic than ketones (pK_a about 20). The stabilising ester resonance

is partially lost in the ester enolate. This is the major reason why ketones are more acidic than esters (pK_a about 25).

SUMMARISE	Deprotonation of β-ketoester **A** with LiOEt gives the resonance stabilised lithium enolate **D-2a**. Of the two possible pairs of geometric enolates, this enolate is preferred as it is less crowded and has internal coordination to the lithium counter-ion. The carbonyl oxygen of the ester group is the more electron rich, and consequently better at coordinating the lithium counter-ion than the oxygen atom in –OEt. The preferred enolate **D-2a** has (*Z*)-stereochemistry.

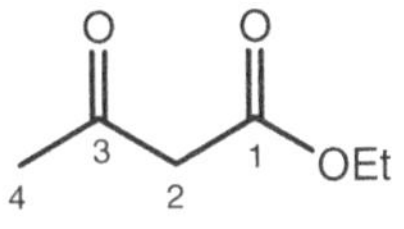

A: $C_6H_{10}O_3$

Y: $C_6H_9O_3Li$

A

CH_3COR $\xrightleftharpoons{\text{base}}$ $[\ominus CH_2COR \longleftrightarrow CH_2=CO\ominus R]$

CH_3CO_2R $\xrightleftharpoons{\text{base}}$ $[\ominus CH_2CO_2R \longleftrightarrow CH_2=CO\ominus OR]$

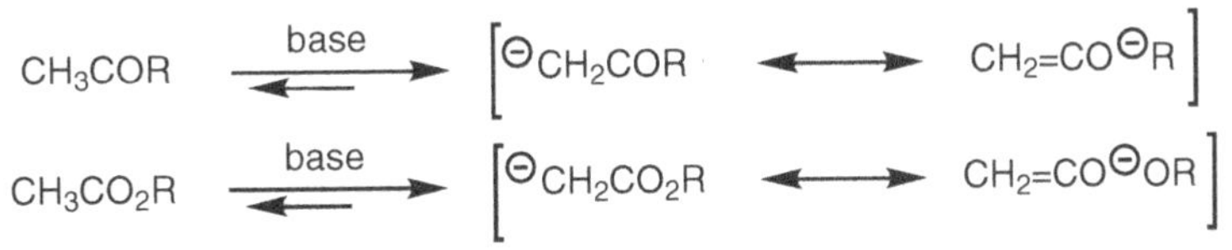

Formation of an enolate by deprotonation of a carbonyl derivative

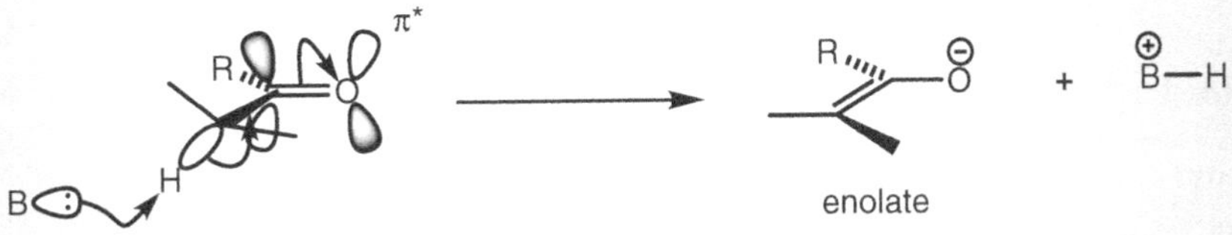

π^*

enolate

Removal of a conformationally acidic proton – HOMO: C–H σ-bond; LUMO: π^* of C=O

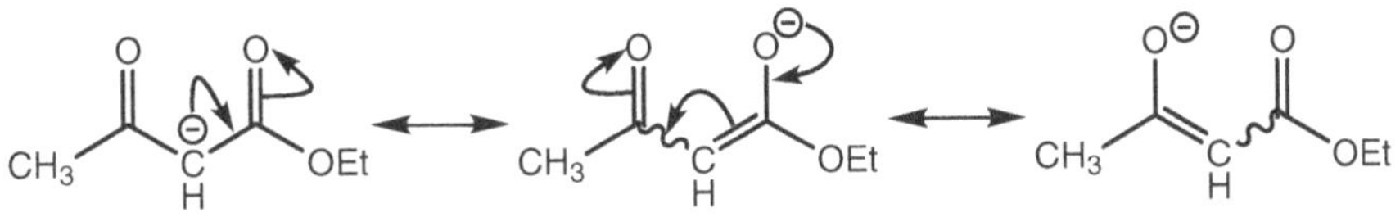

B $\longleftrightarrow$ **C** $\longleftrightarrow$ **D** (E)- or (Z)-?

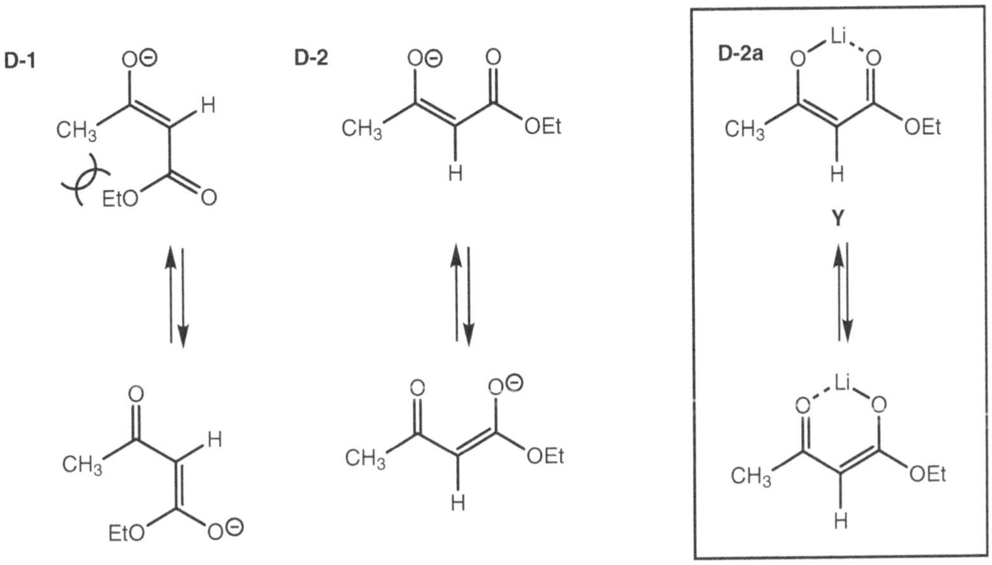

D-1

D-2

D-2a

Y

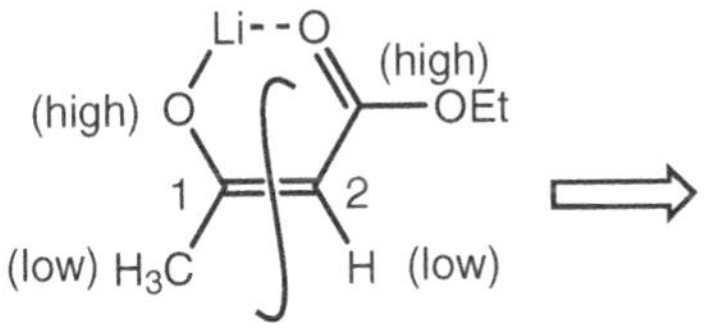

Position	at C(1)		at C(2)	
Group	OLi	CH₃	CO₂Et	H
First atom	**O**	**C**	**C**	**H**
Priority order	High	Low	High	Low

D-2a is **Y**; this enolate has a (Z)-configuration

2. Treatment of the enantiomerically pure ester (S)-A with a solution of NaOEt (in EtOH) resulted in a decrease in optical rotation. Draw a mechanism to account for this observation.

(S)-A

DRAW the STRUCTURE of A in three dimensions

A is a chiral ester, with one chiral centre adjacent to the ester group.

CONSIDER the REAGENTS and OUTLINE a MECHANISM

Ethoxide ion is a base and a nucleophile. A C–H bond next to CO_2R is acidic; the carbanion is stabilised by resonance with the ester group. For this enolisation, the best orbital interaction occurs when bonding C–H orbital overlaps with the antibonding π^*-orbital of the C=O bond. *The pK_a of ethanol is about 16 and of an ester is about 25. There is only a very small amount of enolate present at equilibrium, but this is enough to act as a reactive intermediate in this reaction. Nucleophilic addition of ethoxide to the ester group would make a tetrahedral intermediate which would then eliminate ethoxide; overall no apparent structural change would take place and the chirality of A would be unaffected.*

DRAW out a possible ANSWER and predict the major PRODUCT(S)

The best conformers for easy removal of a proton from **A** are (S)-**A-1** and (S)-**A-2**; both the accessible and give (E)- and (Z)-enolates respectively. These enolates are planar and re-protonation can occur equally on either the top or bottom face, forming racemic **A** (with zero optical rotation) from each enolate. Repeated enolisation and re-protonation results in complete racemisation.

It is important to note that the substituents at C_2 of an enolate are assigned relative priorities as normal using the Cahn-Ingold-Prelog rules (see page 16). However, by convention, for C_1 the highest priority is always given to O-metal, irrespective of the other substituent.

SUMMARISE

Deprotonation of optically pure ester results in the formation of an achiral planar enolate. The ability to rotate the plane of plane-polarised light has been lost at this stage. Re-protonation to re-form the more stable C=O group can occur equally on both faces of the enolate and gives equal amounts of both enantiomers of esters **A** (racemic mixture). Repeated deprotonation and re-protonation of ester (S)-**A** leads to racemisation.

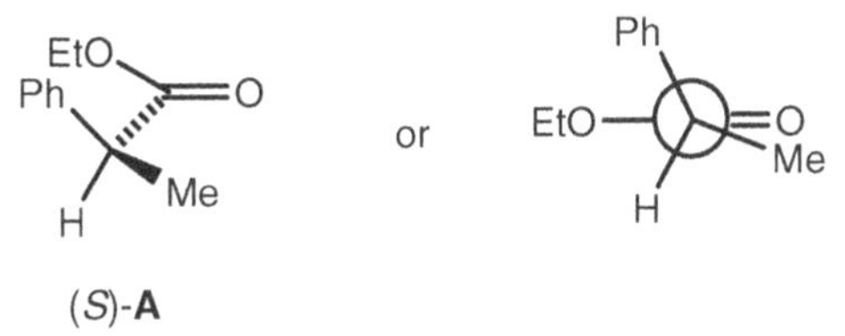

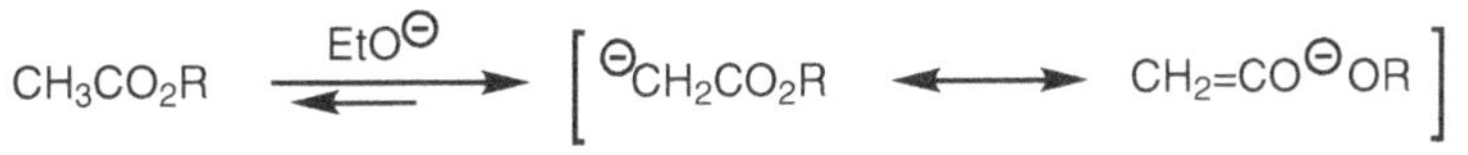

(S)-**A**

$$CH_3CO_2R \xrightarrow{EtO^\ominus} \left[{}^\ominus CH_2CO_2R \longleftrightarrow CH_2=CO^\ominus OR \right]$$

Formation of an enolate by deprotonation of an ester

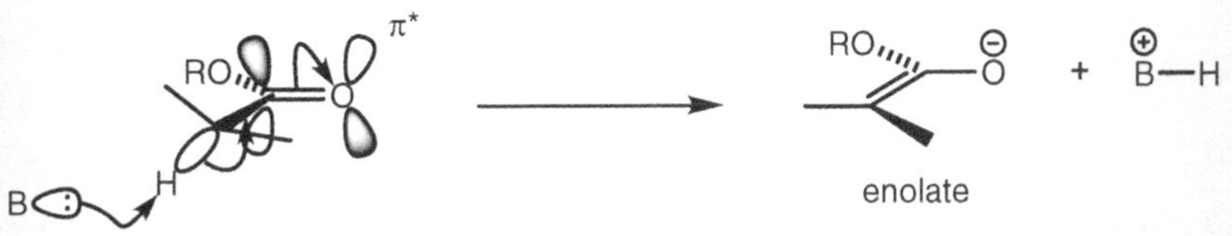

enolate

Removal of the more conformationally acidic proton – HOMO: C–H σ-bond; LUMO: π* of C=O

Enolisation:

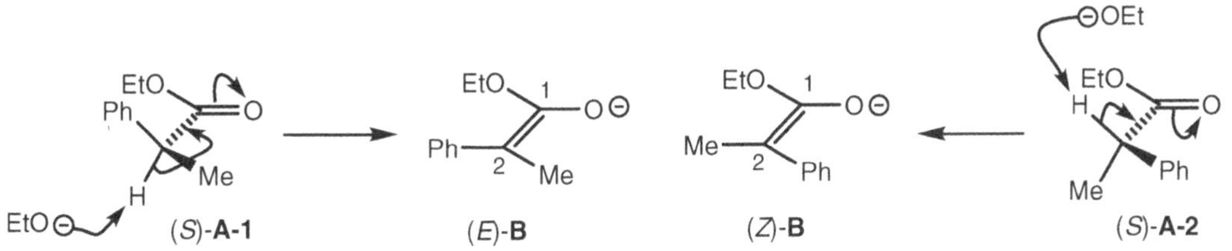

Re-protonation:

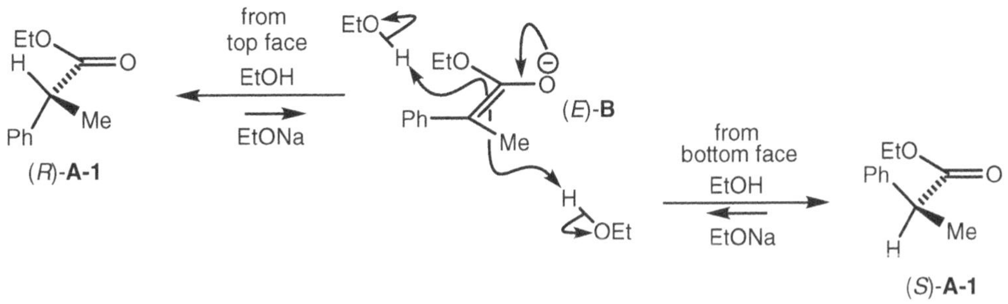

Similarly,

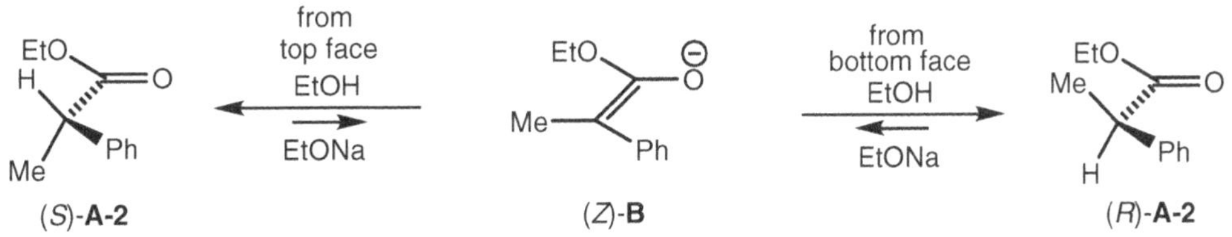

Overall,

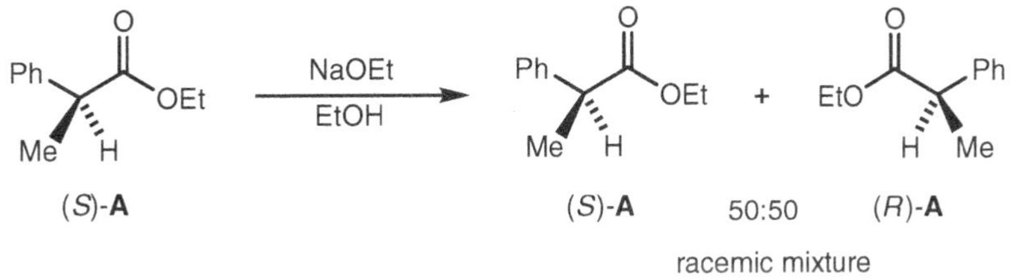

(S)-**A**   (S)-**A**   50:50   (R)-**A**

racemic mixture

3. The conversion of *cis*-1-decalone **A** to its more stable isomer *trans*-1-decalone **B** can be achieved by addition of NaOEt in EtOH. Draw a mechanism to account for this isomerisation and comment on the possible driving force for this reaction.

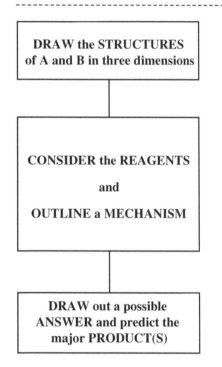

cis-**A**   *trans*-**B**

DRAW the STRUCTURES of A and B in three dimensions	A is a *cis*-decalin and is flexible with two major conformers, *cis*-**A-1** and *cis*-**A-2**. *trans*-Decalin **B** has one major conformer. **A** and **B** are epimers at C_6. *Both A and B are chiral. This is an epimerisation because only one of the two chiral centres in cis-A has changed in the reaction.*
CONSIDER the REAGENTS **and** **OUTLINE a MECHANISM**	The ethoxide ion is a base and a nucleophile. The protons H_a, H_b and H_c are all acidic, being next to C=O, and suitable orientations for deprotonation can be achieved (for H_b in *cis*-**A-1** and for H_a and H_c in *cis*-**A-2**). Enolisation by loss of H_a followed by re-protonation could cause epimerisation at C_6. *Addition of ethoxide to C=O is reversible and no change would be observed. Enolisation to C_2 and re-protonation would also give no detectable change.*
DRAW out a possible ANSWER and predict the major PRODUCT(S)	Enolate **Y** is formed by loss of H_a from *cis*-**A-2**. Protonation of **Y** from the top face gives back *cis*-**A-2** but from below gives *trans*-**B**. Because *trans*-**B** has less internal steric hindrance than *cis*-**A**, *trans*-**B** is more stable and will predominate at equilibrium. *In cis-decalins, the additional steric crowding arises from close approch of the hydrogens inside the curved carbon framework and from extra gauche interactions across the cis-bridge.*

SUMMARISE	Ketone *trans*-**B** can be formed by epimerisation of *cis*-**A** by simple deprotonation using NaOEt (to form enolate **Y**) and stereoselective re-protonation (using EtOH). This reaction is under thermodynamic control, and consequently favours formation of the more stable diastereoisomeric ketone *trans*-**B**.

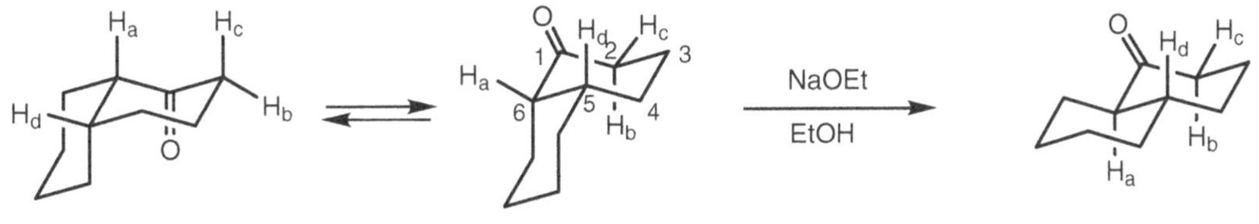

cis-**A-2**   *cis*-**A-1**   *trans*-**B**

$$CH_3COR \xrightleftharpoons{EtO^{\ominus}} \left[{}^{\ominus}CH_2COR \longleftrightarrow CH_2{=}CO^{\ominus}R \right]$$

Formation of an enolate by deprotonation of a ketone

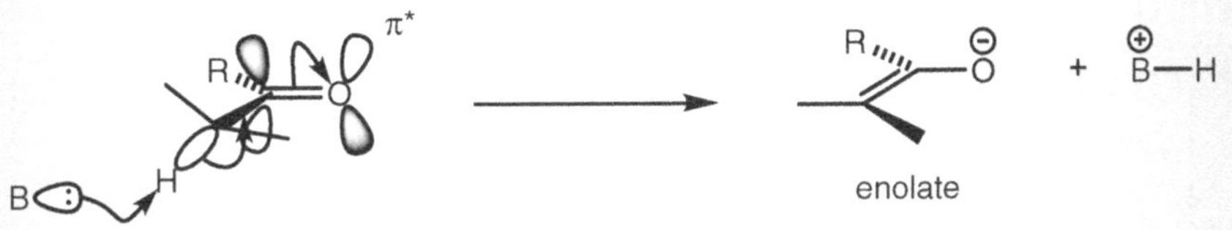

enolate

Removal of the most conformationally acidic proton – HOMO: C–H σ-bond; LUMO: π* of C=O

Only enolisation across C_1–C_6 will change the chirality at C_6. Best conformer is *cis*-**A-2**.

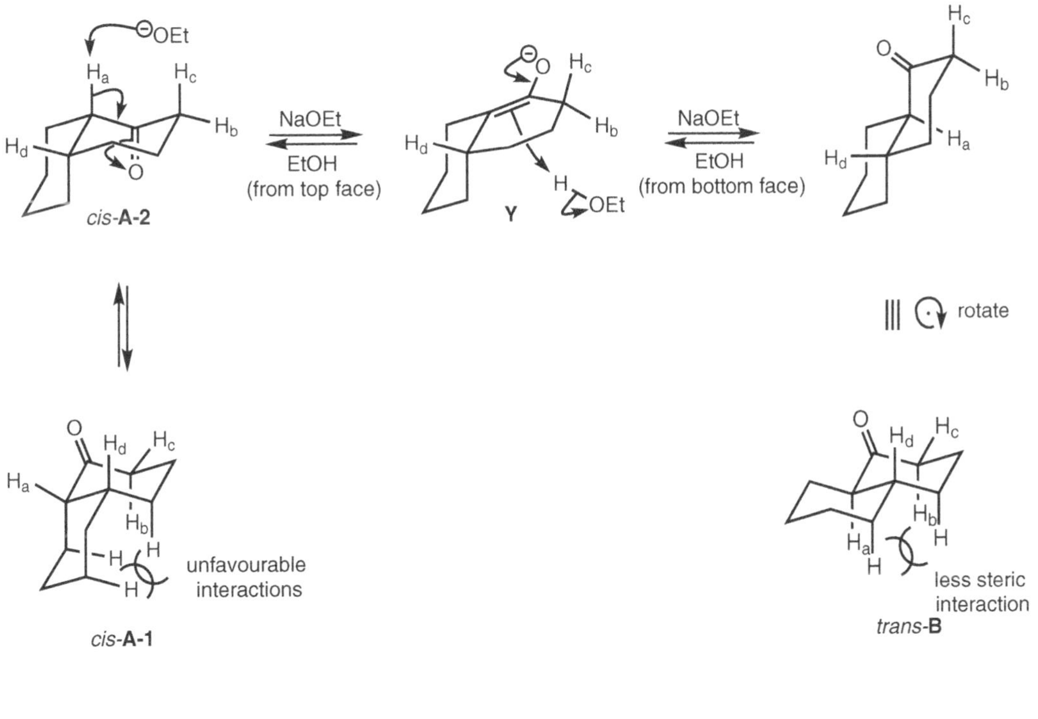

cis-**A-2**

NaOEt / EtOH (from top face)

NaOEt / EtOH (from bottom face)

Y

rotate

cis-**A-1**   unfavourable interactions

trans-**B**   less steric interaction

4. Stereoselective methylation of ester *syn*-**A** can be achieved by addition of LDA in THF and then treating the resulting enolate with methyl iodide. This gave a mixture of diastereoisomeric esters *syn*- and *anti*-**B** in the ratio 84:16. Rationalise the formation of the major diastereoisomer.

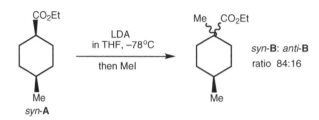

DRAW the STRUCTURES of A and B in three dimensions	Both **A** and **B** are achiral. *syn*-**A** has two major conformations, each with one axial and one equatorial substituent. *syn*- and *anti*-**B** are drawn here in their most stable conformations, with only one axial substituent. *Remember that ring-flip in cyclohexanes changes all axial substituents into equatorial ones (and vice versa), and that larger substituents are more stable in equatorial positions. An Me group on cyclohexane behaves as though it is a* larger *group than an ester group.*
CONSIDER the REAGENTS and OUTLINE a MECHANISM	LDA is a very strong base (the pK_a for *i*-Pr_2NH is about 35). THF is a non-protic solvent. A proton adjacent to the ester group (pK_a about 25) will be removed via a six-membered ring transition state to give a lithium enolate **Y**. In the second stage, the electron-rich enolate acts as a nucleophile in an S_N2 reaction with MeI to make the α-methylated ester. *Ionisation of LDA is difficult in a non-protic solvent. Coordination of LDA with the carbonyl compound in a chair-like arrangement lowers the activation energy for deprotonation.*
DRAW out a possible ANSWER and predict the major PRODUCT(S)	MeI approaches the enolate **Y** from above or below the plane, where the enolate's HOMO has its highest electron density. Approach from the top left-hand side (as drawn) is hindered by the two axial hydrogens, whereas approach from the lower right-hand side is more accessible; thus *syn*-**B** is formed faster than *anti*-**B**. *Starting with the less stable conformer, syn-**A**-2, the same enolate **Y** can be formed by deprotonation and a ring flip .*
SUMMARISE	Deprotonation of ester *syn*-**A** and diastereoselective methylation of the resulting racemic enolate **Y** (on the less hindered pseudo-equatorial face) leads to the required major diastereoisomeric ester, *syn*-**B**.

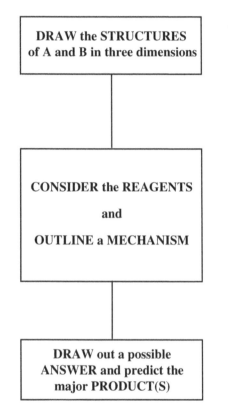

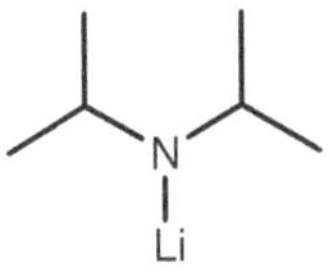

LDA
Lithium
Diisopropyla**m**ide

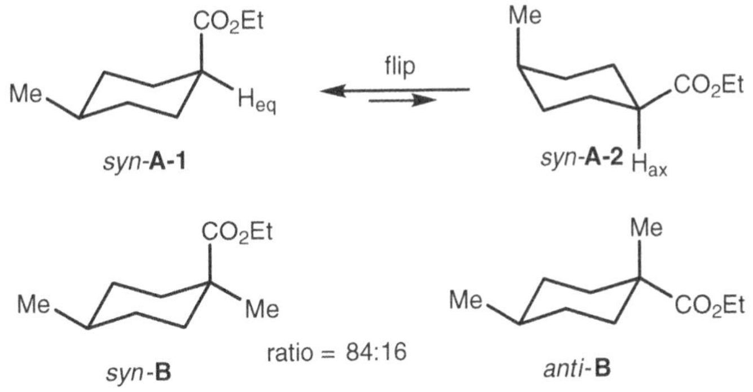

syn-**A-1** ⇌ (flip) syn-**A-2** H_{ax}

syn-**B**   ratio = 84:16   anti-**B**

$$CH_3CO_2R \xrightarrow{LiN(Pr)_2} \left[{}^{\ominus}CH_2CO_2R \longleftrightarrow CH_2=CO^{\ominus}OR \right] \xrightarrow[S_N2]{MeI} MeCH_2CO_2R$$

Formation of an enolate using LDA

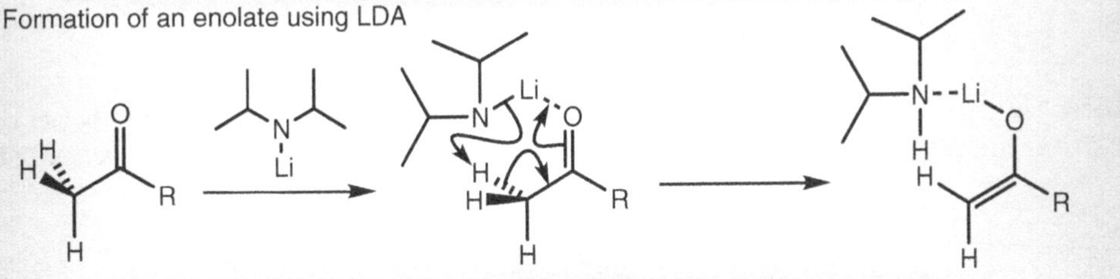

six-membered cyclic chair transition state

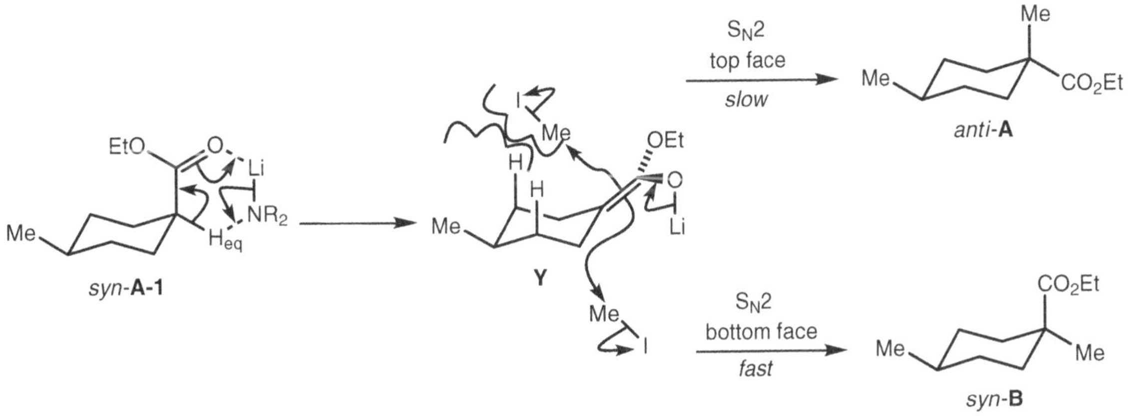

syn-**A-1**

Y

S_N2
top face
slow

anti-**A**

S_N2
bottom face
fast

syn-**B**

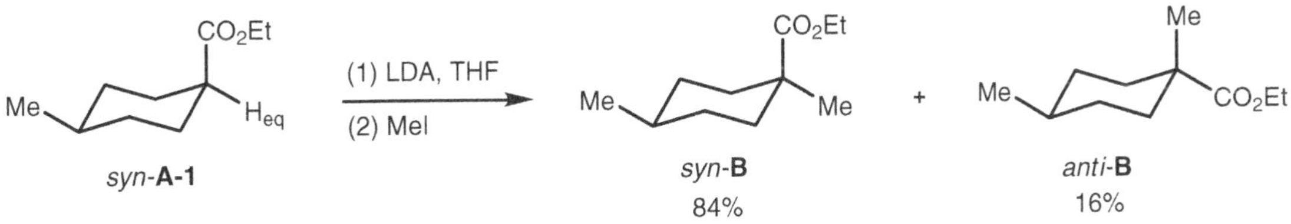

syn-**A-1** $\xrightarrow[\text{(2) MeI}]{\text{(1) LDA, THF}}$ syn-**B** 84% + anti-**B** 16%

Additional problems for Chapter 6

1. Draw the structure and assign the configuration of the major enolate formed by deprotonation of the ketone **A**, amide **B** and ester **C**.

A   **B**   **C**

2. Account for the formation of the isomeric silyl enol ethers **D** and **F** under the following conditions and assign their enol configuration.

D   **E**   **F**

3. The oxazoline *syn*-**G** was epimerised using NaOEt in EtOH to give the more stable diastereoisomeric oxazoline *anti*-**H**. Draw a mechanism to account for this observation.

single enantiomer
syn-**G**

single enantiomer
anti-**H**

4. Deprotonation of the ketone (±)-**I** with LDA in THF, followed by methylation of the resulting enolate with MeI gave a single diastereoisomeric adduct (±)-**J**. Draw a mechanism to account for the stereochemical outcome of this reaction.

I   **J**

5. Treatment of ketone (±)-**J** with NaOEt (in ethanol) resulted in epimerisation of only one of the two stereocentres adjacent to the carbonyl to give the isomeric (±)-**K**. Account for this overall change in stereochemistry, and predict what product might be obtained if the reaction was repeated with NaOEt in EtOD.

J   **K**

Answers: 1. **A** forms the (*Z*)-enolate by removal of the more acidic hydrogen adjacent to Cl; **B** forms the (*Z*)-enolate due to minimisation of strain with adjacent N−Me$_A$ bond; **C** forms the (*E*)-enolate due to the presence of the smaller OMe group; note: O⁻ has a higher priority than OMe (see page 112); 2. (*E*)-**D** is formed under thermodynamic control; reversible conditions allow formation of the more stable double bond. (*E*)-**F** is formed under kinetic control by removing the more acidic, least hindered proton using a hindered base; this reaction is irreversible; 3. Deprotonation of **G** with methoxide and stereorandom re-protonation with EtOH allows formation of the more stable isomer **H** (see page 114); 4. Axial methylation via a chair TS; 5. Formation of the more stable stereoisomer by deprotonation and re-protonation (see page 114). Using EtOD would replace both Hs adjacent to C=O in **K** – the stereochemistry would remain unchanged.

7 Polar Rearrangements

The aim of this chapter is to develop an understanding of the stereochemical features involved within stepwise rearrangement processes.
  At the end of this chapter, you should be able to:

- Draw and compare the mechanisms of simple rearrangement processes involving [1,2]-alkyl or aryl shifts.
- Predict and explain the stereochemical consequence of a given rearrangement.
- Explain the conformational influence at the migration origin and terminus.

Background

Reactions involving C–C bond rearrangement allow rapid assembly of complex carbon and heteroatomic skeletons that otherwise would be difficult to achieve using traditional synthetic chemistry. Broadly, there are two main classes:

(1) *Pericyclic rearrangements*: these are orbitally controlled rearrangements which can be achieved under thermal or photochemical assistance. These concerted rearrangements are not discussed in this text.

(2) *Polar rearrangements*: these rearrangements are stepwise and generally involve a [1,2]-alkyl/aryl shift.

These polar shifts are promoted by an electron-deficient migration terminus (usually a σ^*- or an empty p-orbital) and an electron-rich migration origin. The rearrangement starts by donation of the nucleophilic σ-bond (its HOMO) at the migration origin into the unoccupied, coplanar antibonding orbital (LUMO, σ^*- or empty p-orbital) at the migration terminus.

Stepwise polar rearrangements

These reactions proceed in two ways:

(1) Using an electron-deficient migration terminus (e.g. in carbocation-assisted [1,2]-alkyl shifts). Stereochemically, **retention** of configuration of the migrating substituent R* is observed. If the reaction is concerted there will be **inversion** at both the migration origin and terminus.

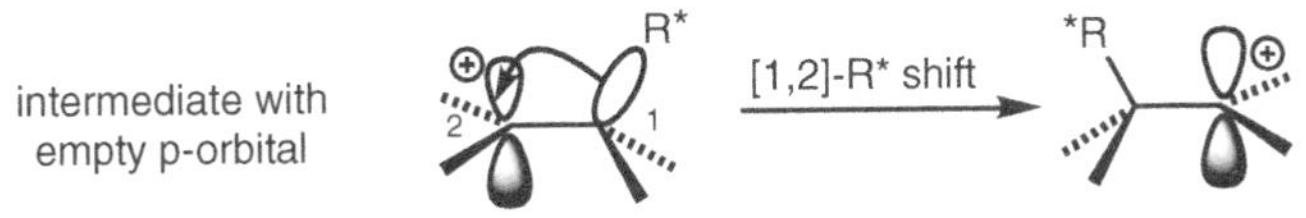

or, if concerted

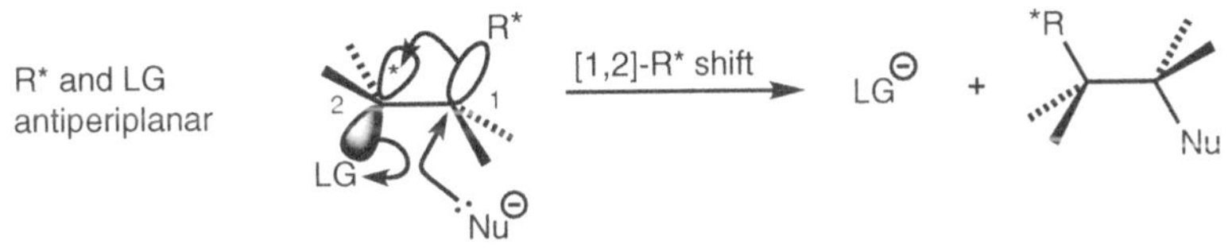

(2) Using an electron-rich migration origin (e.g. in alkoxide-assisted [1,2]-alkyl shifts). Characteristically, retention of configuration of the migrating substituent R* occurs, and in addition, inversion of configuration at the migration terminus is observed, e.g.

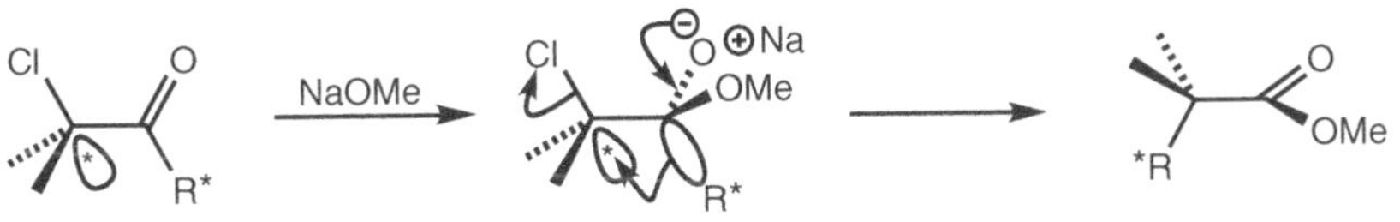

Examples of these rearrangements include:

- the Beckmann rearrangement of oximes to amides
- the Hofmann rearrangement of amides to amines
- peroxy rearrangements, such as the Baeyer-Villiger oxidation of ketones to esters and the oxidation of boranes by H_2O_2
- carbocation induced shifts, such as the pinacol rearrangement.

1. Draw the structure of the two isomeric products, **B** and **B′**, derived from the thermal Beckmann rearrangement of *N*-tosyl imines **A** and **A′** (〜〜 represents undefined stereochemistry).

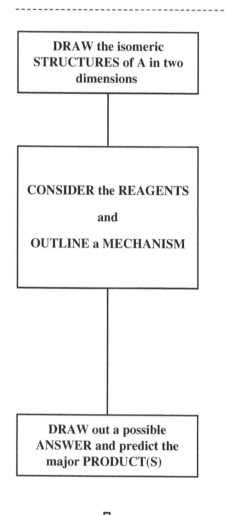

$$\text{heat} \atop H_2O$$

B (C_8H_9NO) + **B′** (C_8H_9NO)

DRAW the isomeric STRUCTURES of A in two dimensions	**A** has two geometric isomers due to the lack of rotation about the C=N bond. *The C and N of C=N are both sp² hybridised; the structure is trigonal and flat, similar to an alkene with the N non-bonded pair occupying a stereochemical position like an H on ethene. In (E)-/(Z)- assignments, a lone pair of electrons has lower priority than any atom.*
CONSIDER the REAGENTS and **OUTLINE a MECHANISM**	From **A** to **B**, OTs is lost and OH is gained; ⁻OTs is a good leaving group, so the OH probably comes from water. Loss of ⁻OTs leaves behind a partially positive centre, which is the electron-deficient migration terminus. An alkyl or aryl group migrates with its bonding electrons to the developing partial positive centre as the ⁻OTs leaves. This reaction is a **Beckmann rearrangement**; it is stereospecific and the group *anti*-to the leaving group always migrates. *Other Beckmann-type processes include the rearrangement of oximes (R₂C=NOH) and related compounds, such as R₂C=NOSO₂R, R₂C=NHal and R₂C=NN₂⁺. The anti-stereospecificity arises from the necessary orbital overlap between the bonding orbital of the migrating group (HOMO) and the anti-bonding orbital at the migration terminus (see **W**).*
DRAW out a possible ANSWER and predict the major PRODUCT(S)	The group *anti*- to the leaving group migrates: Ph for (*E*)-**A** and Me for (*Z*)-**A′**. The intermediates **X** and **Y** are hydrolysed to give enolic structures which rapidly tautomerise into the more stable and more familiar amide groups. *Water adds to the carbon of the multiple bond in **X** and **Y** and not to the nitrogen. This nitrogen atom (N⁺) is not electron-deficient as it already has four bonded pairs (like NH₄⁺). Addition to carbon atom, as shown, is the preferred pathway.*
SUMMARISE	Thermal rearrangement of *N*-tosyl imine (*E*)-**A** and (*Z*)-**A′** gives the amides **B** and **B′** respectively. This process is stereospecific; Ph (in **A**) and Me (in **A′**) migration must occur via concerted loss of tosylate. Hydrolysis of the intermediate nitrilium ions (**X** and **Y**) gives the required amides **B** and **B′**.

A and **A′** are $C_{15}H_{15}NO_3S$ (or C_8H_8NOTs); **B** and **B′** are C_8H_9NO; ^-OTs is a good leaving group

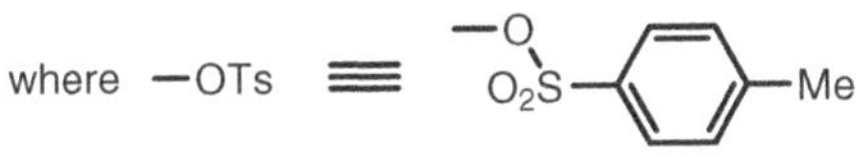

has (E)-stereochemistry

has (Z)-stereochemistry

where $-OTs \equiv$

$$R_2C=NOTs \xrightarrow{H_2O} RNHCOR + TsOH$$

Beckmann rearrangement: conversion of an oxime to an amide

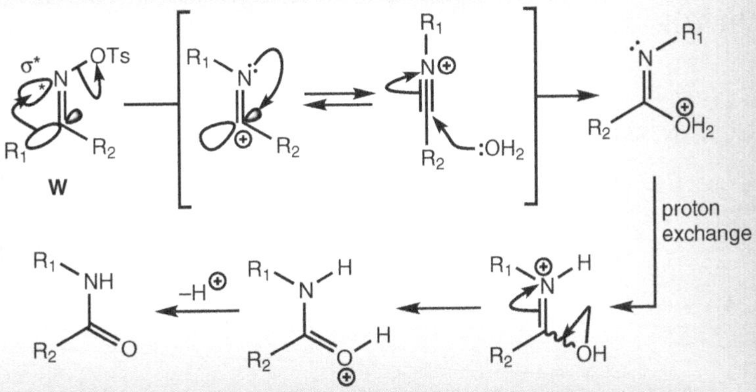

Migration occurs *anti*- to the leaving group with retention of configuration in R_1

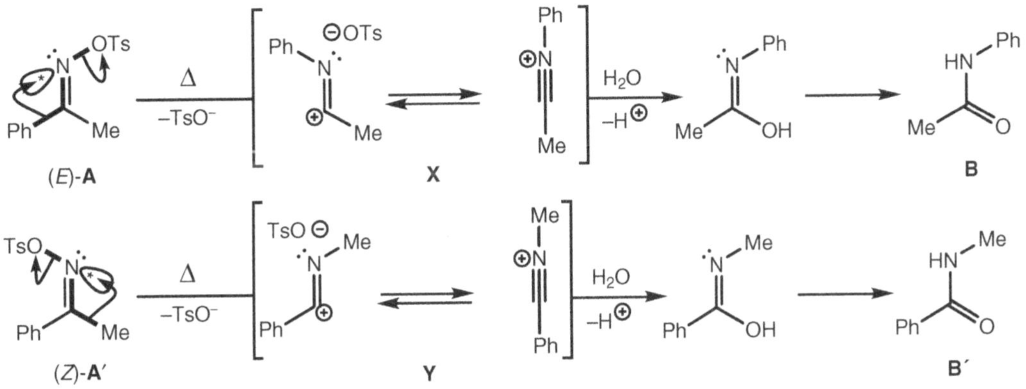

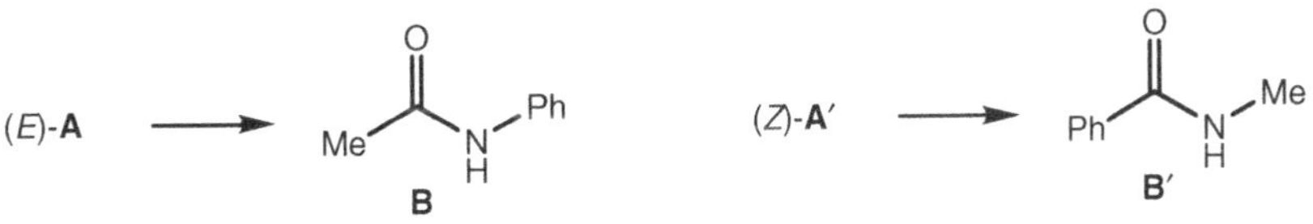

2. Predict the stereochemical outcome for the Hofmann rearrangement of amide (*S*)-**A** to give compound **B**.

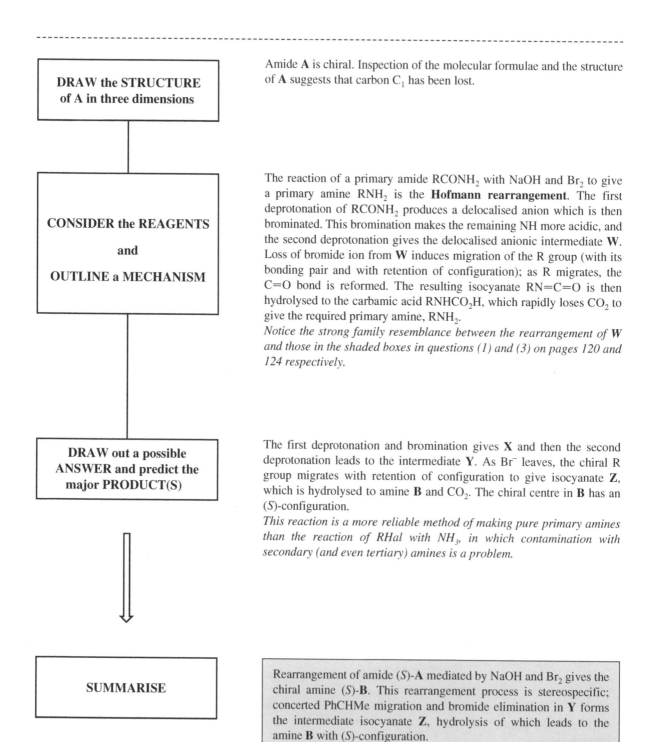

PhCH(Me)CONH₂, (*S*)-**A** $\xrightarrow{\text{NaOH, Br}_2}$ **B** (C₈H₁₁N)

DRAW the STRUCTURE of A in three dimensions

Amide **A** is chiral. Inspection of the molecular formulae and the structure of **A** suggests that carbon C₁ has been lost.

CONSIDER the REAGENTS

and

OUTLINE a MECHANISM

The reaction of a primary amide RCONH₂ with NaOH and Br₂ to give a primary amine RNH₂ is the **Hofmann rearrangement**. The first deprotonation of RCONH₂ produces a delocalised anion which is then brominated. This bromination makes the remaining NH more acidic, and the second deprotonation gives the delocalised anionic intermediate **W**. Loss of bromide ion from **W** induces migration of the R group (with its bonding pair and with retention of configuration); as R migrates, the C=O bond is reformed. The resulting isocyanate RN=C=O is then hydrolysed to the carbamic acid RNHCO₂H, which rapidly loses CO₂ to give the required primary amine, RNH₂.

*Notice the strong family resemblance between the rearrangement of **W** and those in the shaded boxes in questions (1) and (3) on pages 120 and 124 respectively.*

DRAW out a possible ANSWER and predict the major PRODUCT(S)

The first deprotonation and bromination gives **X** and then the second deprotonation leads to the intermediate **Y**. As Br⁻ leaves, the chiral R group migrates with retention of configuration to give isocyanate **Z**, which is hydrolysed to amine **B** and CO₂. The chiral centre in **B** has an (*S*)-configuration.

This reaction is a more reliable method of making pure primary amines than the reaction of RHal with NH₃, in which contamination with secondary (and even tertiary) amines is a problem.

SUMMARISE

Rearrangement of amide (*S*)-**A** mediated by NaOH and Br₂ gives the chiral amine (*S*)-**B**. This rearrangement process is stereospecific; concerted PhCHMe migration and bromide elimination in **Y** forms the intermediate isocyanate **Z**, hydrolysis of which leads to the amine **B** with (*S*)-configuration.

A: $C_9H_{11}NO$ and **B**: $C_8H_{11}N$ – one carbon and one oxygen atom has been lost; maybe C_1?

Group	$CONH_2$	H	Me	Ph
First atom	C	**H**	C	C
Second atom	**O,O,N**		**H,H,H**	**C,C,C**
Priorities	1	4	3	2

(S)-

(S)-**A**

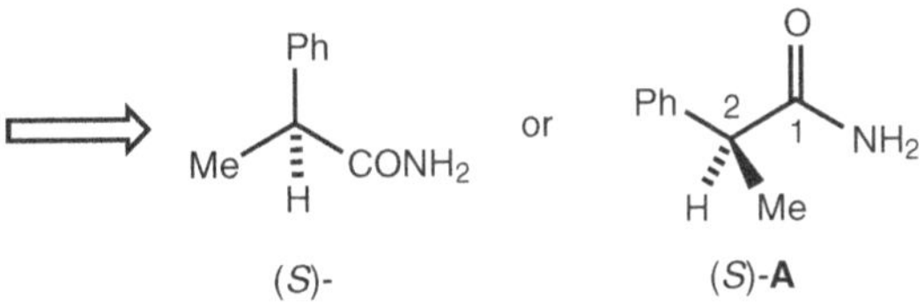

$$RCONH_2 \xrightarrow{\text{NaOH, Br}_2} \left[RN=C=O \right] \xrightarrow{H_2O} RNH_2 + CO_2$$

Hofmann rearrangement: conversion of an amide to an amine and CO_2

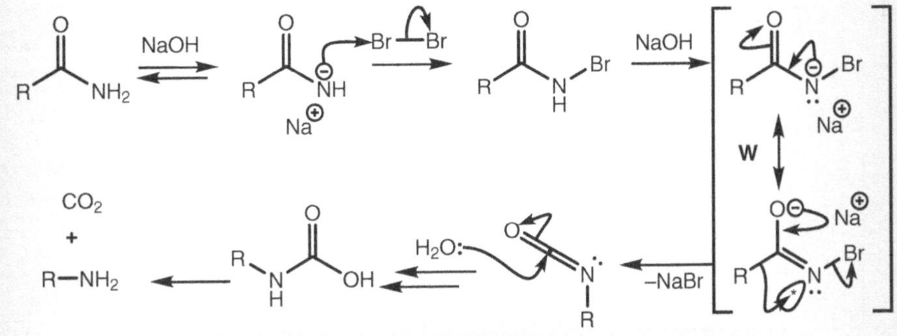

Migration occurs *anti-* to the leaving group with retention of configuration in R

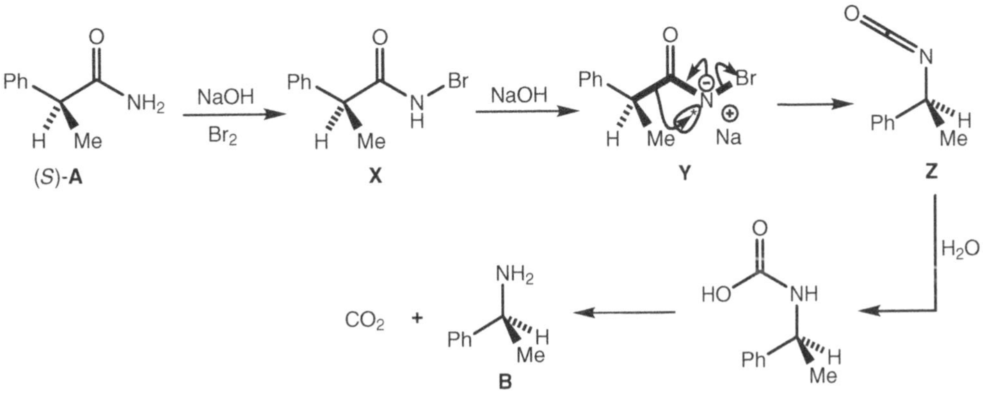

(S)-**A**   **X**   **Y**   **Z**

B

Group	NH_2	H	Me	Ph
First atom	**N**	**H**	C	C
Second atom			**H,H,H**	**C,C,C**
Priorities	1	4	3	2

(S)-

(S)-**B** $(C_8H_{11}N)$

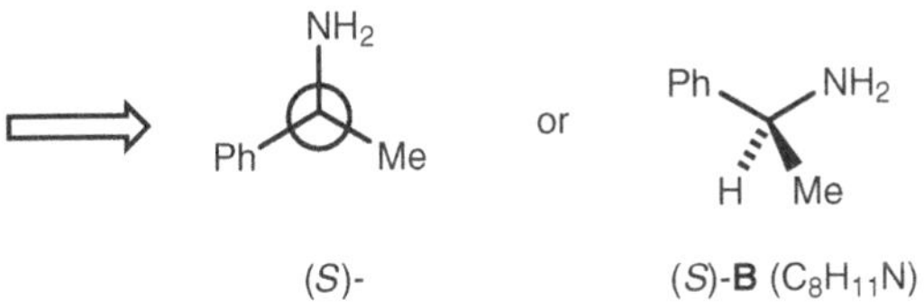

3. Draw a mechanism and account for the stereochemistry of the product alcohol (*S*)-**B**, derived from the ketone (*S*)-**A** using a peroxy rearrangement reaction.

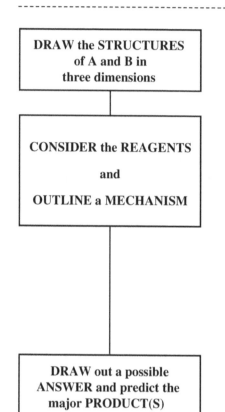

(*S*)-**A** Stage 1 **Z** Stage 2 (*S*)-**B**

DRAW the STRUCTURES of A and B in three dimensions	Ketone **A** is converted into alcohol **B** with retention of configuration. Comparing the structures suggests that the C_2–C_3 bond is broken.
CONSIDER the REAGENTS and OUTLINE a MECHANISM	CF_3CO_3H is a per-acid which reacts with $NaHCO_3$ to give CF_3CO_3Na, H_2O and CO_2. In stage 1, CF_3CO_3Na will add to the ketone to give the tetrahedral peroxy intermediate, which rearranges (with loss of the good leaving group $CF_3CO_2^-$) to give an ester. This reaction is a **Baeyer-Villiger rearrangement**. The resultant ester is hydrolysed by aqueous NaOH in stage 2. *The lengthening of the weak O–O bond in the peroxy adduct W creates an electron-deficient centre on O to which an electron-rich R group can migrate with its bonding pair (with retention of configuration in R). At the same time, a non-bonded pair on O⁻ re-forms a C=O. By rotation about the RR′CO⁻–OO bond in W, either R or R′ groups could be brought into the correct anti orientation for rearrangement.*
DRAW out a possible ANSWER and predict the major PRODUCT(S)	Two tetrahedral conformers **X** and **Y** can be formed from **A** with either MeCHD or Me antiperiplanar, as shown. Rearrangement is faster when R is electron rich and large (the latter making it more likely that R will be antiperiplanar to the leaving group across the O–O bond), and so MeCHD will migrate faster than Me. Ester (*S*)-**Z** is formed with retention of configuration. Hydrolysis of (*S*)-**Z** in aqueous NaOH via the tetrahedral intermediate mechanism ($B_{AC}2$) gives (*S*)-**B**. *An Me group generally does not migrate because it is better at stabilising the developing transition state by hyperconjugation. If Me did migrate, a methyl ester would be formed. On hydrolysis this would give the wrong products: methanol and the chiral acid (S)-2-deuteriopropanoic acid, MeCDHCO₂H.*
SUMMARISE	Per-acid addition to the ketone (*S*)-**A**, and subsequent hydrolysis of the intermediate ester gives the required alcohol (*S*)-**B**. Stereospecific oxidation of (*S*)-**A** proceeds with retention of configuration; MeCDH has a greater migratory aptitude than Me.

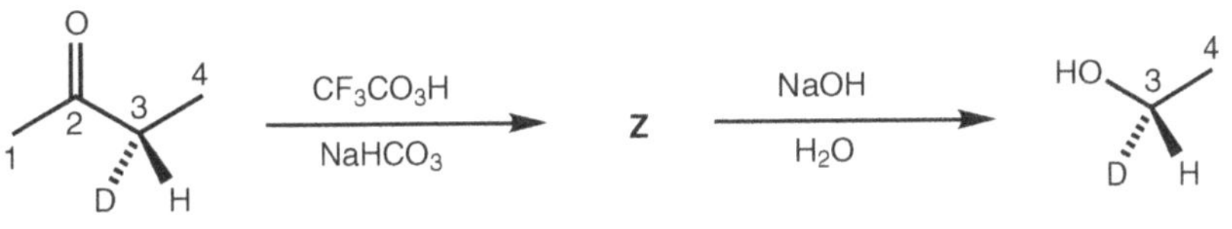

A is C₄H₇DO and **B** is C₂H₅DO; *two carbon atoms are lost – due to C₂–C₃ fission?*

$$RR'C=O \xrightarrow{R''CO_3H} RCO_2R' \xrightarrow{H_2O} RCO_2H + R'OH$$

A Baeyer-Villiger rearrangement: conversion of a ketone to an ester using a per-acid

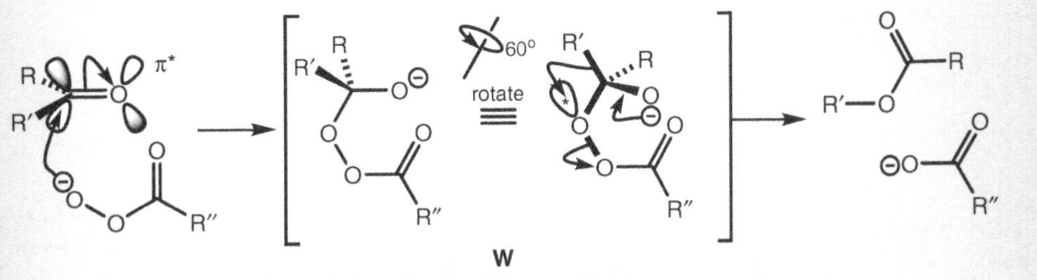

Migration occurs *anti-* to the leaving group with retention of configuration in R′.

Stage 1 – Baeyer-Villiger: (*S*)-**A** to intermediate **Z**.

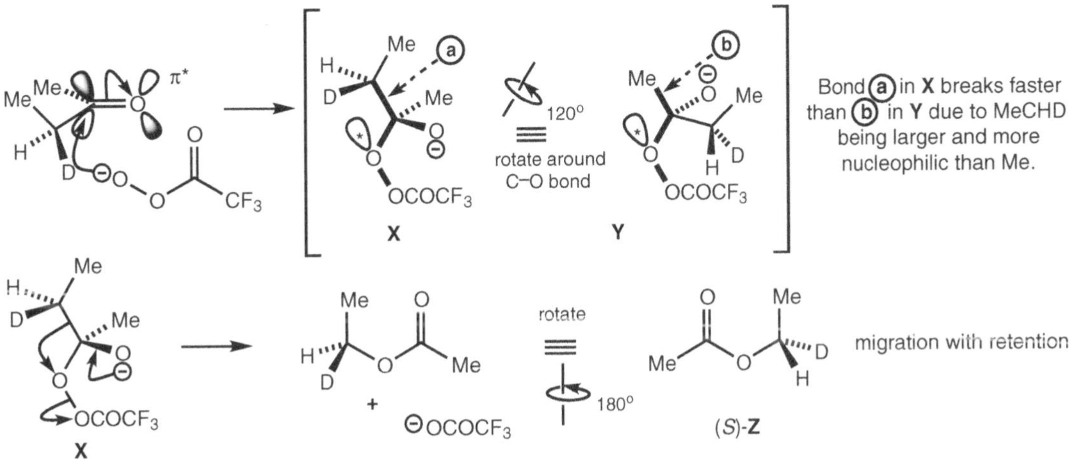

Bond **a** in **X** breaks faster than **b** in **Y** due to MeCHD being larger and more nucleophilic than Me.

migration with retention

Stage 2 – Ester hydrolysis: chiral centre not affected: conversion of (*S*)-**Z** to (*S*)-**B**.

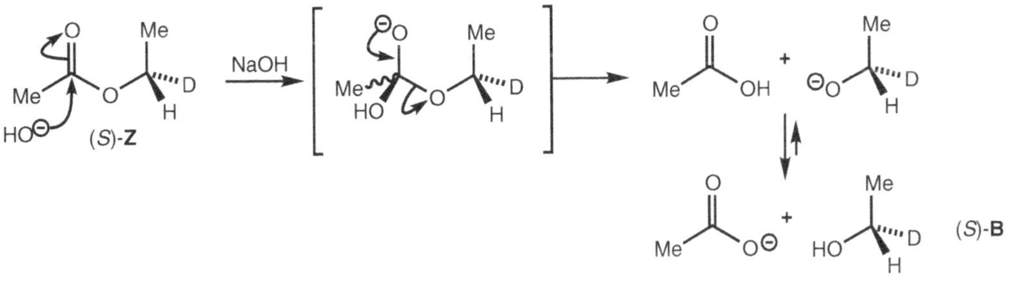

4. Predict the stereochemistry of the secondary alcohol (±)-**B**, derived from the oxidative hydroboration of 1-methyl cyclohexene, **A** with HBR₂ (where R = CMe₂CHMe₂).

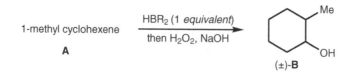

1-methyl cyclohexene

A

HBR₂ (1 *equivalent*)
then H₂O₂, NaOH

(±)-**B**

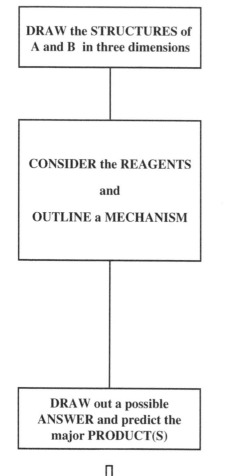

DRAW the STRUCTURES of A and B in three dimensions

A is an alkene, **B** is a secondary alcohol. **B** can have either *syn*- or *anti*-stereochemistry, each having two chair conformers.
Since A and all the reagents are achiral, either syn- or anti-B would be produced in racemic form.

CONSIDER the REAGENTS and OUTLINE a MECHANISM

Boranes add to alkenes to give alkylboranes, which are oxidised to alcohols by alkaline hydrogen peroxide.
When the borane is oxidised, the H₂O₂ is reduced (to H₂O).
Borane addition to C=C is *syn*- and regioselective; the boron atom adds to the less hindered end. HOO⁻ adds to the empty p-orbital of the new electron-deficient boron, and the borate anion **X** rearranges with loss of hydroxide ion and migration of an alkyl group from boron to oxygen. Overlap of the C–B σ-bond (HOMO) with the O–O σ* (LUMO) ensures that the migration occurs with retention of configuration. Addition of ⁻OH to the boron followed by breakage of the CO–B bond gives the alcohol product.
Notice that the boron switches from six outer-shell electrons to eight and back to six electrons as, for example, ⁻OH adds to the boron and the alkoxide is lost.

DRAW out a possible ANSWER and predict the major PRODUCT(S)

Starting with **A** and following this mechanism through gives the *anti*-borane C. Oxidation and rearrangement (with retention of configuration in the migrating cyclohexyl group) gives **D**, which is hydrolysed to the alcohol **B**.
B has the same stereochemistry as the intermediates C and D, and its stereochemistry is set up in the first (addition step). Using the large –CMe₂CHMe₂ (thexyl) substituents on the borane reagent ensures good regioselectivity, and also poor competition with the required migration of the cyclohexyl group.

SUMMARISE

Oxidative hydroboration of alkene **A** gives the required alcohol *anti*-**B**. Regioselective hydroboration of **A** proceeds via *syn*-stereoselective addition of H and BR₂ across the C=C bond to give racemic *anti*-2-methylcyclohexyl-dithexylborane; BR₂ adds to the less hindered and more nucleophilic end of the alkene. Oxidative exchange of BR₂ for an OH group occurs stereospecifically with retention of configuration, by addition of alkaline hydrogen peroxide and elimination of hydroxide.

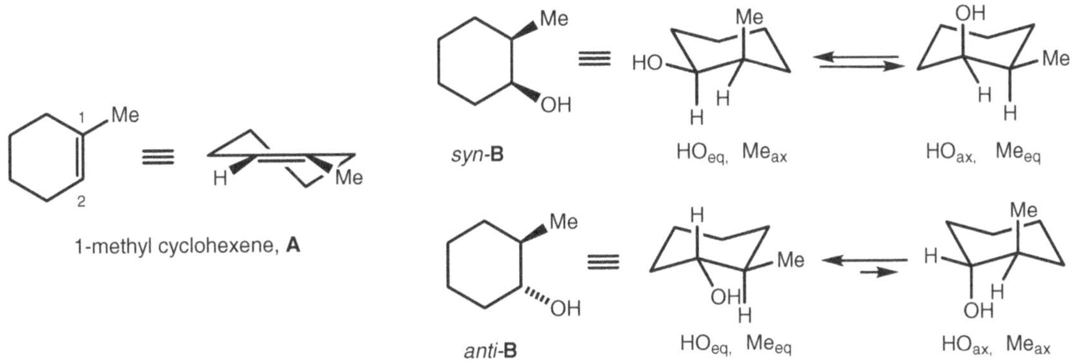

1-methyl cyclohexene, **A**

syn-**B**

HO$_{eq}$, Me$_{ax}$ HO$_{ax}$, Me$_{eq}$

anti-**B**

HO$_{eq}$, Me$_{eq}$ HO$_{ax}$, Me$_{ax}$

$$R_2C=CR_2 \xrightarrow{\text{HBR}'_2} R_2CHCR_2BR'_2 \xrightarrow[\text{H}_2\text{O}_2]{\text{NaOH}} R_2CHCR_2OH$$

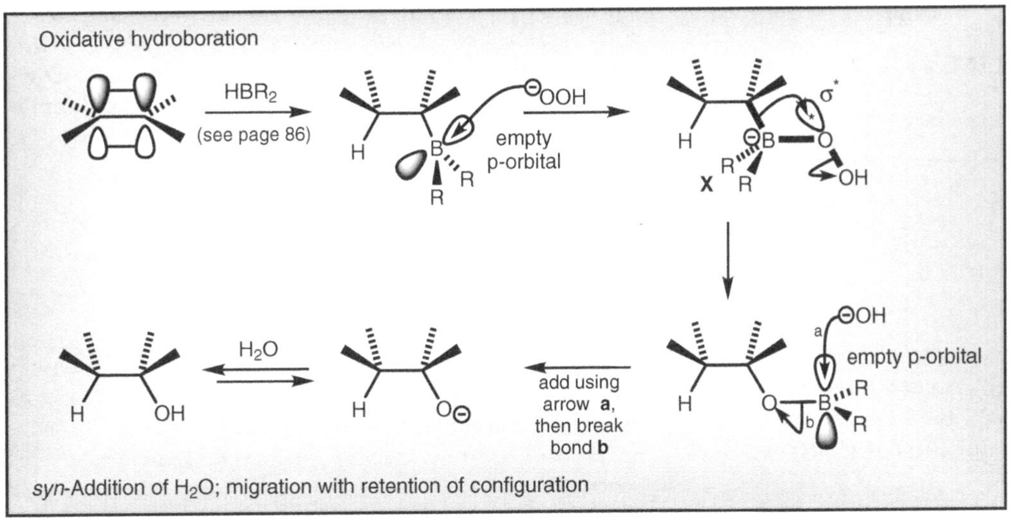

Oxidative hydroboration

HBR$_2$ (see page 86)

empty p-orbital

empty p-orbital

add using arrow **a**, then break bond **b**

H$_2$O

syn-Addition of H$_2$O; migration with retention of configuration

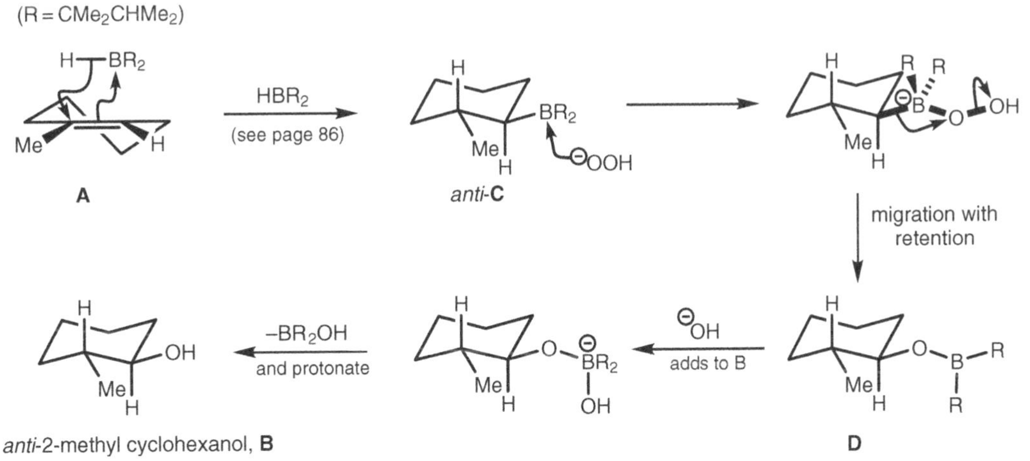

(R = CMe$_2$CHMe$_2$)

HBR$_2$ (see page 86)

A   *anti*-**C**

migration with retention

$-$BR$_2$OH and protonate

$^{\ominus}$OH adds to B

D

anti-2-methyl cyclohexanol, **B**

5. The major products from the acid-catalysed reactions of diols **A** and (±)-**C** are shown below. Explain these different results.

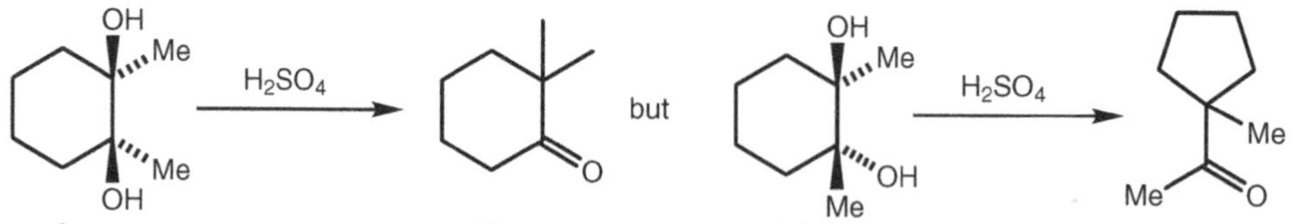

OH / Me / Me / OH
A → H$_2$SO$_4$ → **B** but OH / Me / OH / Me (±)-**C** → H$_2$SO$_4$ → Me / Me / O **D**

DRAW the STRUCTURES of A and C in three dimensions	**A** and **C** are diastereoisomeric; **A** is a *syn*-diol and **C** is an *anti*-diol. Each has two chair conformers. In each reaction H$_2$O has been lost, but a different rearrangement of the carbon skeleton has occurred.

CONSIDER the REAGENTS

and

OUTLINE a MECHANISM

Acid will protonate an OH, making it a good leaving group. In the **pinacol rearrangement** of 1,2-diols, a group migrates to the electron-deficient centre caused by the loss of the water, at the same time, a new carbonyl group is formed from the second OH. The leaving group and the migrating group must lie antiperiplanar to each other (see page 119). *Two reactions which may compete with the pinacol rearrangement are elimination of water (to give an enol and then a C=O group), and epoxidation, in which the second OH adds to the electron-deficient centre. Neither of these reactions causes a rearrangement of the carbon skeleton.*

DRAW out a possible ANSWER and predict the major PRODUCT(S)

Both **A-1** and **A-2** have an antiperiplanar diaxial relationship between OH and Me groups. After protonation of an axial OH in either **A-1** or **A-2**, water is lost and a methyl group migrates as the C=O bond is formed, giving the resonance-stabilised cation **X**. Deprotonation of **X** gives ketone **B**.

In **C-1**, the leaving group is antiperiplanar only to H or OH. Possible reactions are the formation of either the enol (from loss of H and H$_2$O) or the epoxide (from participation of the OH), but neither of these reactions would give **D**.

In **C-2**, the leaving group is antiperiplanar only to two of the ring C–C bonds. A rearrangement involving one of these bonds leads to a ring contraction (from a six-membered ring to a five-membered ring) and the formation of the resonance-stabilised cation **Y**. Loss of a proton from **Y** gives **D**.

*A similar rearrangement of the other ring C–C bond would lead to an unstable primary carbocation and evidently this pathway is not favoured. Ring contraction reactions are also possible for **A**, but require more energy than the pinacol reaction.*

SUMMARISE

Acid-catalysed rearrangement of diols **A** and **C** gave ketones **B** and **D** respectively. Elimination of H$_2$O is assisted by donation of adjacent antiperiplanar σ-bonding electrons. This promotes Me group migration in diol **A** and ring contraction in diol **C**.

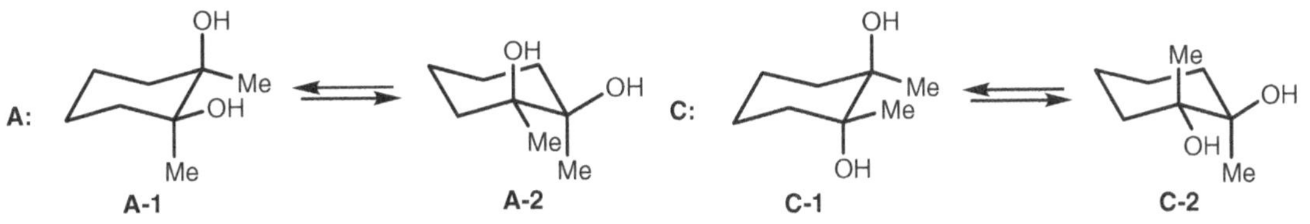

A: A-1 ⇌ A-2   C: C-1 ⇌ C-2

A/C $(C_8H_{16}O_2)$ $\xrightarrow{-H_2O}$ B/D $(C_8H_{14}O)$

Pinacol rearrangement: 1,2-diol to ketone/aldehyde

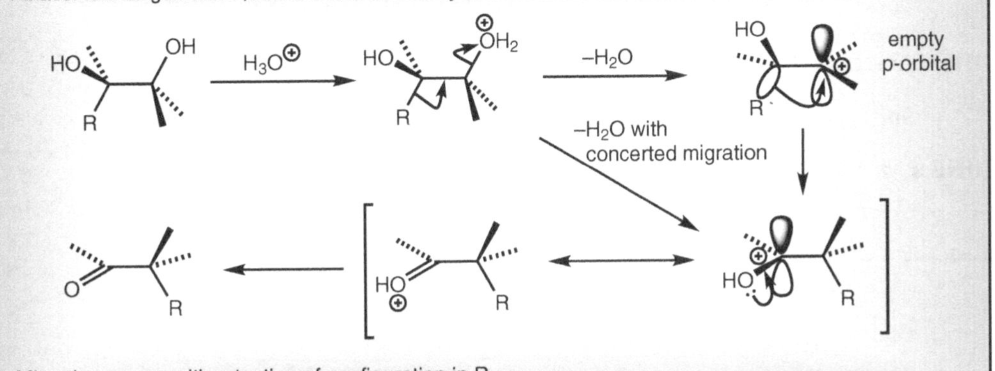

$R_2C(OH)C(OH)CR_2$ $\xrightarrow{H_3O^{\oplus}}$ $R_3C(R)C=O$ + H_2O

$\xrightarrow{H_3O^{\oplus}}$

$\xrightarrow{-H_2O}$

empty p-orbital

−H₂O with concerted migration

Migration occurs with retention of configuration in R

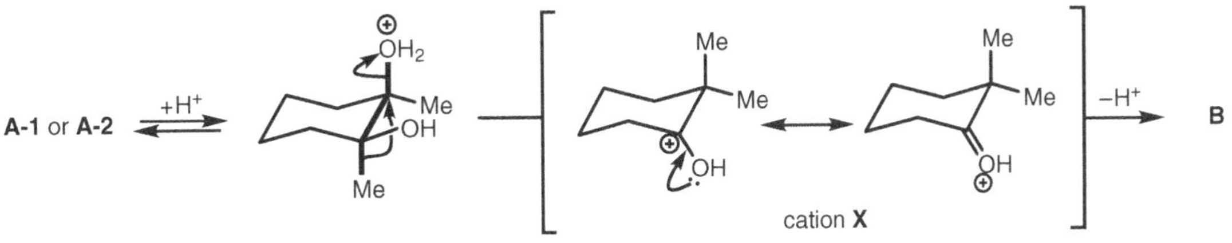

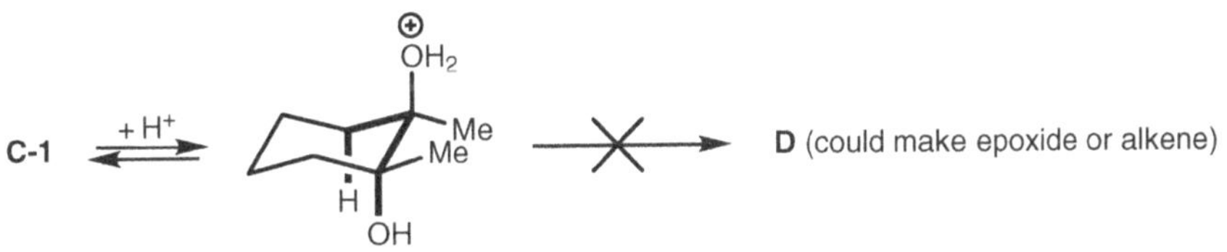

A-1 or A-2 $\underset{}{\overset{+H^+}{\rightleftharpoons}}$ [cation X] $\xrightarrow{-H^+}$ B

C-1 $\overset{+H^+}{\rightleftharpoons}$ ⤫ D (could make epoxide or alkene)

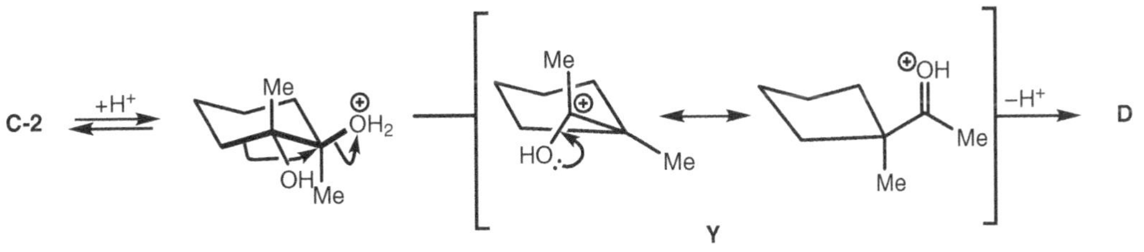

C-2 $\overset{+H^+}{\rightleftharpoons}$ [Y] $\xrightarrow{-H^+}$ D

6. Treatment of racemic chloro-alcohols **A** under basic conditions can give three different products **B**, **C** and **D**. Determine which products can be made from each stereoisomer of **A**.

DRAW the STRUCTURE of A in three dimensions

Chloro-alcohol **A** has *syn-* and *anti*-stereoisomers, and each of these has two chair conformers. **B** is a ketone, **C** is an epoxide and **D** is an aldehyde. Making **D** will require a skeletal rearrangement. **B**, **C** and **D** are all formed by loss of HCl from **A**.

CONSIDER the REAGENTS

and

OUTLINE a MECHANISM

Overall loss of HCl suggests that the NaOH is acting as a base to remove the most acidic proton, which is on the OH group.

The transition state for epoxidation requires the $^-$O$^-$ and the Cl to be anti-periplanar (overlap between non-bonded pair on O and the σ* of C–Cl). Similarly, for migration of H or R, H/R and Cl must be antiperiplanar (overlap between H/R–C's bonding pair and the σ* of C–Cl). As the H/R migrates, a C=O bond is formed.

Compare these migrations with those in the Baeyer-Villiger rearrangement on page 124).

DRAW out a possible ANSWER and predict the major PRODUCT(S)

The only antiperiplanar arrangement for **1,2-substituents** on a cyclo-hexane ring is axial, axial. The O$^-$ and Cl can achieve this in the ax, ax conformer of *anti*-**A** to give epoxide **C**, as can the H and Cl in the eq, ax conformer of *syn*-**A** to give the ketone **B**. In the other two conformers, the Cl is equatorial and has no antiperiplanar **substituents**; but there are two C–C ring bonds antiperiplanar to the C–Cl bond. Migration of these ring bonds can take place in conjunction with the loss of Cl$^-$ and favour-able formation of a strong new C=O bond in **D**.

The energy released on making the stable C=O can be considered as 'paying for' the high-energy ring contraction process.

SUMMARISE

Of the four possible stereoisomers of racemic **A**, only the two diastereo-isomers need to be considered. Base-mediated rearrangement of **A** leads to three products; each diastereoisomer of **A** (and associated ring flip conformer) has a preferred elimination pathway governed by the group positioned antiperiplanar to the chloride leaving group; *anti*-**A**$_{eq,eq}$ and *syn*-**A**$_{ax,eq}$ favour ring contraction to form the alde-hyde **D**, whereas *syn*-**A**$_{eq,ax}$ prefers a hydride shift to give the ketone **B**. Direct displacement using an alkoxide to give the epoxide **C** can only occur via conformer *anti*-**A**$_{ax,ax}$.

anti-**A**   syn-**A**

A (HO$_{eq}$,Cl$_{eq}$)   **A** (HO$_{ax}$,Cl$_{ax}$)   **A** (HO$_{ax}$,Cl$_{eq}$)   **A** (HO$_{eq}$,Cl$_{ax}$)

relative energy $\mathbf{A_{eq,eq} < A_{ax,ax}}$ relative energy $\mathbf{A_{ax,eq} \approx A_{eq,ax}}$

Epoxidation or rearrangement

Antiperiplanar requirements: $^{\ominus}$O–C–C–Cl or R–C–C–Cl

anti-$\mathbf{A_{eq,eq}}$ anti-$\mathbf{A_{ax,ax}}$ syn-$\mathbf{A_{ax,eq}}$ syn-$\mathbf{A_{eq,ax}}$

NaOH   NaOH   NaOH   NaOH

D   **C**   **D**   **B**

7. Base-assisted fragmentation of racemic **A** gives two unsaturated ketones **B**. Predict which diastereoisomer of **A** is responsible for each isomer of ketone **B** (〰 represents undefined stereochemistry).

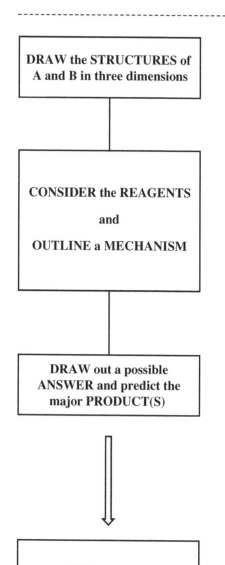

DRAW the STRUCTURES of A and B in three dimensions

Hydroxy-tosylate **A** has a decalin skeleton. Its two diastereoisomers are the *trans*-decalin, **A-1**, and the *cis*-decalin, **A-2**. The cyclodecenone, **B**, with its C=C bond has two (*E/Z*) geometric isomers.
*Flexible cis-decalins can 'ring-flip' between two double chair conformations, but the more rigid trans-decalins cannot (see page 11). Formation of **B** involves the loss of TsOH from **A**.*

CONSIDER the REAGENTS and OUTLINE a MECHANISM

NaOEt will deprotonate the OH, and the TsO⁻ is good leaving group. Asuming that C_5 of **A** becomes the carbonyl group of **B**, counting round the structures suggests that C_5–C_6 is broken and that the new C=C bond is at C_1–C_6. If these bond makings and breakings are concerted, this is a **Grob fragmentation** and requires an antiperiplanar arrangement of ⁻O–C–C–C–OTs in **W**.
The ring σ C–C (HOMO) overlaps with the σ* (LUMO) of the departing OTs. The three non-bonded pairs on O⁻ can easily provide a pair anti to the breaking C–C. Notice again the formation of a very stable C=O bond in this reaction, as in questions (5) and (6) (see pages 128 and 130 respectively).

DRAW out a possible ANSWER and predict the major PRODUCT(S)

A-1 has the required antiperiplanar stereochemistry. The positions of the C_1/C_6 H atoms hardly change, corresponding to an (*E*)- C=C bond in **B**. Similarly, with **A-2**$_{eq}$ the C_1/C_6 H atoms smoothly convert to the (*Z*)-isomer. Conformer **A-2**$_{ax}$ has no antiperiplanar ⁻O–C–C–C–OTs arrangement, however, *trans*-diaxial elimination is possible.

SUMMARISE

Base-assisted elimination from the two diastereoisomeric tosylates **A-1** and **A-2** leads to two isomeric ketones (*E*)- and (*Z*)-**B** respectively. The elimination of tosylate occurs through antiperiplanar C–C participation (with some help from the neighbouring alkoxide) to give the stable C=O bond.

There are two diastereoisomers of **A** $\xrightarrow[?]{- \text{TsOH}}$ two diastereoisomers of **B**

HOCH(R)CH(R)CH$_2$OTs $\xrightarrow{\text{NaOH}}$ RCHO + RCH=CH$_2$ + TsOH

Grob fragmentation

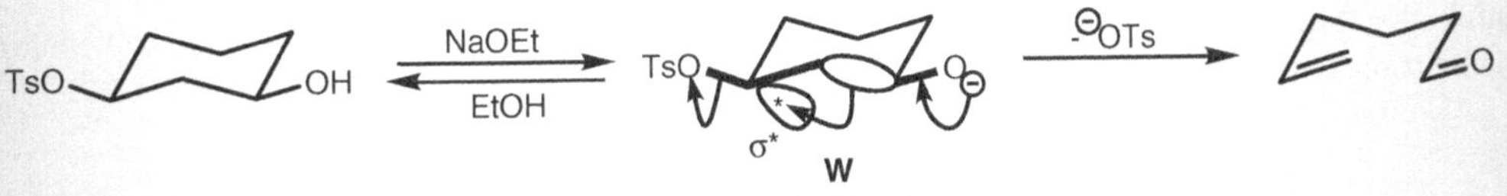

Antiperiplanar elimination involving the σ-electrons of the adjacent carbon–carbon bond

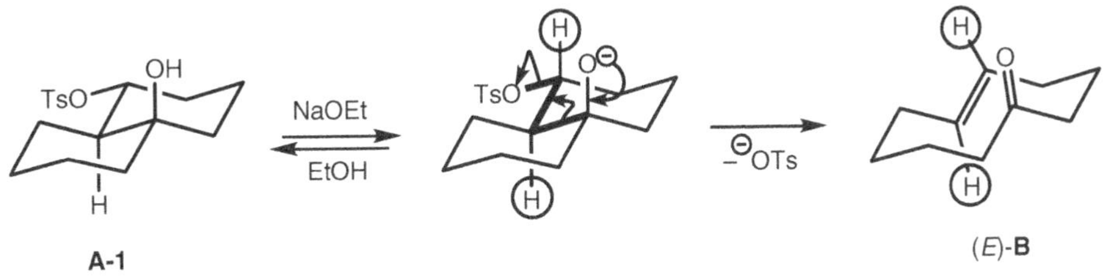

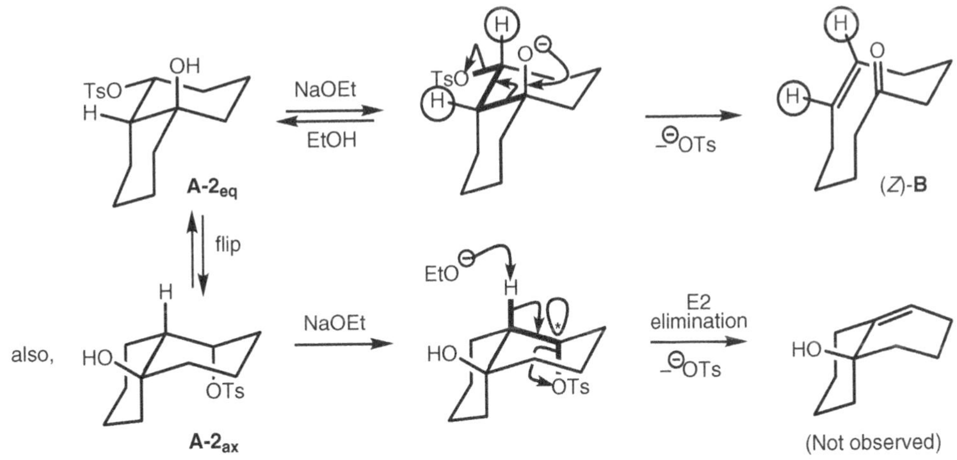

Additional problems for Chapter 7

1. Treatment of one of the two possible stereoisomers of oxime **A** with sulfuric acid gave the amide **B**. Predict which stereoisomer of the oxime **A** was responsible for this reaction.

$$(E)\text{- or }(Z)\text{-}\mathbf{A} \qquad \xrightarrow{\text{H}_2\text{SO}_4} \qquad \mathbf{B}$$

2. Treatment of the primary amine (*R*)-**C** with nitrous acid leads to the tertiary alcohol (*R*)-**D**. Rationalise the formation of this alcohol, and account for the stereochemical outcome of this rearrangement.

$$(R)\text{-}\mathbf{C} \qquad \xrightarrow{\text{HNO}_2} \qquad (R)\text{-}\mathbf{D}$$

3. Rationalise the formation of ketone **F** derived from the addition of nitrous acid to the amino alcohol **E**. Draw a mechanism and assign the configuration present in **F**.

single enantiomer
E
$$\xrightarrow[\text{AcOH}]{\text{HNO}_2}$$
F

4. The lactone (±)-**H** was formed by addition of trifluoroperacetic acid (CF$_3$CO$_3$H) to a solution of ketone (±)-**G**. Draw a mechanism to account for this reaction and rationalise the stereochemical outcome of this oxidative insertion.

$$\mathbf{G} \qquad \xrightarrow[\text{CH}_2\text{Cl}_2]{\text{CF}_3\text{CO}_3\text{H}} \qquad \mathbf{H}$$

5. Skeletal rearrangement of α-chloro ketone (±)-**I** with NaOMe gave the ester (±)-**J**. Account for the stereochemical outcome of this transformation.

$$\mathbf{I} \qquad \xrightarrow[\text{Et}_2\text{O}]{\text{NaOMe}} \qquad \mathbf{J}$$

Answers: 1. (*Z*)-**A** (see page 120); 2. HNO$_2$ converts the NH$_2$ into $^+$N$_2$; the iso-butyl substituent migrates with retention of configuration to give a tertiary carbocation. This is trapped with water to give the tertiary alcohol (*R*)-**D**; 3. HNO$_2$ converts the NH$_2$ into an $^+$N$_2$ group. Methyl group migration *via* an S$_N$2-type process gives the required inversion of configuration at the migration terminus. The new stereocentre has an (*S*)-configuration; 4. This is a Baeyer-Villiger reaction (see page 124); 5. This is a Favorskii-type reaction. Addition of methoxide to the ketone; the resulting alkoxide assists Ph migration to displace the adjacent chloride substituent with overall inversion of configuration.

Index

Symbols and Abbreviations

General orbitals

·· or ⬭:	lone pair/non-bonded pair (n) orbital (HOMO)
⬭	bonding (σ) orbital (HOMO)
⬭*	antibonding (σ*) orbital (LUMO)
⬮⊕	non-bonding (LUMO) orbital of a carbocation
⬮⊖	non-bonding (HOMO) orbital of anion
	π-bond (HOMO) orbital
	π*-antibonding (LUMO) orbital

Substituents

Ph	Phenyl
Me	Methyl
Et	Ethyl
i-Pr or *iso*-Pr	*iso*-Propyl
t-Bu or *tert*-Bu	*tert*-Butyl
OAc	Acetate
OTs	Tosylate
LDA	LiN(*i*-Pr)$_2$
THF	O⬠

General chemical bonds

∿∿	stereochemistry random/unidentifed/ non-specified
⎯⎯	bond in plane of paper
▬	emphasised bond in plane of paper
◢	bond projecting forward in front of paper
┈┈┈	bond projecting backward behind paper
‑ ‑ ‑ ‑	partly formed bond (e.g. in transition state)

Chemical arrows

≡	equivalent to (drawn differently)
↻	by rotation in the plane of the paper (z-axis)
⟳	by rotation in the vertical axis (x-axis)
⟳	by rotation in the horizontal axis (y-axis)
⟳	by rotation about a given axis
⟶	goes to
⟶⟶	goes to in several stages
⇌	equilibrium
[⟷]	resonance structures

General chemical symbols

+H$^+$, −H$^+$	proton exchange
Ⓢ Ⓜ Ⓛ	small, medium, large substituents
E$^+$	generalised electrophile
Nu$^-$	generalised nucleophile
LG	generalised leaving group
EWG	generalised electron withdrawing group
HA	generalised acid
B :	generalised base
e$^⊖$	electron (solvated)
R	inactive aliphatic side chain
Ar	inactive aryl side chain
ax	axial substituent
eq	equatorial substituent
(±)-	denotes racemic compound
syn-	substituents that are on the same face of the molecule
anti-	substituents that are on the opposite face of the molecule
δ+	a partial postive charge
δ−	a partial negative charge
⤫	internal (intramolecular) steric congestion
∿	external (intermolecular) steric hinderance

Printed and bound in the UK by
CPI Antony Rowe, Eastbourne

Printed and bound by CPI Group (UK) Ltd, Croydon, CR0 4YY
02/03/2022
03115565-0001